NEC® 2008 NEED TO KNOW

About the Author

K. J. Keller is currently the CAD Designer, Safety & Compliance Director, and Project Coordinator for one of New England's most prominent electrical contractors. With over 30 years in the construction industry, she has been a journeyman plumber, has worked in both remodeling and new construction, and has extensive experience in residential and commercial general contracting and the plumbing and electrical trades. Ms. Keller is the co-author of McGraw-Hill's *Electrician's Exam Study Guide*.

NEC® 2008 NEED TO KNOW

The 20% of the Code You Need 80% of the Time

621.319
KEL

K. J. Keller

New York   Chicago   San Francisco   Lisbon   London   Madrid
Mexico City   Milan   New Delhi   San Juan   Seoul
Singapore   Sydney   Toronto

Library of Congress Cataloging-in-Publication Data

Keller, K. J. (Kimberley J.)
   NEC 2008 need to know : the 20% of the code you need 80% of the time / K.J. Keller.
   p. cm.
   Includes index.
   ISBN 978-0-07-150845-2 (alk. paper)
   1. Electric engineering--Insurance requirements--Handbooks, manuals, etc. 2. Electric
wiring--Standards--Handbooks, manuals, etc. 3. Electric wiring--Insurance requirements--
Handbooks, manuals, etc. I. Title.
   TK260.K45 2008
   621.319′240218--dc22

                          2008014432

McGraw-Hill books are available at special quantity discounts to use as premiums and sales
promotions, or for use in corporate training programs. To contact a special sales representa-
tive, please visit the Contact Us page at www.mhprofessional.com.

NEC® 2008 Need to Know

1 2 3 4 5 6 7 8 9 0 DOC/DOC 0 1 4 3 2 1 0 9 8

ISBN 978- 0-07-150845-2
MHID 0-07-150845-7

Sponsoring Editor	Copy Editor	Production Supervisor
Joy Bramble Oehlkers	Wendy Lochner	Pamela A. Pelton
Acquisitions Coordinator	**Proofreader**	**Composition**
Rebecca Behrens	Roger Woodson	Lone Wolf Enterprises, Ltd.
Editorial Supervisor	**Indexer**	**Art Director, Cover**
David E. Fogarty	K. J. Keller	Jeff Weeks
Project Manager		
Virginia Howe		

*Although every effort has been made to make the explanation of the Code accurate, neither the
Publisher nor the Author assumes any liability for damages that may result from the use of the
NEC® 2008 Need To Know.*

Dedication

Electrical planning and installations require a lot of patience, but it's nothing compared to the degree of patience and support I receive from my family. This book is dedicated first to my two wonderful children, Afton and Adam, who show me patience whether I am writing, taking too long in the grocery store, or trying to explain something they probably already know. You are my lighthouse and my shining star, and I love you more than tongue or pen can tell. I would also like to dedicate this book to my mom and dad, Joy and Bob Wallace, who are always here with me, and who have always believed that I can do anything. Through their examples I have learned how to reach for the stars and still keep my head out of the clouds.

Contents

Preface

If you are an apprentice, journeyman, master electrician, designer, estimator, or contractor, then right now you hold in your hand the key to understanding and applying the essential sections of the NEC®. Purchasing this book could very well be the best business decision you could make.

Think about a typical electrical job. There are codes that you have to comply with on a regular basis. The average electrician or contractor is not out installing wiring or equipment in a hazardous or classified location every day or connecting fire pumps, fuel cells, escalators, or hoists. If you were to take your code book and cut out all the articles and sections that only pertain to the work you do day in and day out, you would probably by holding about 20 percent of the overall text in your hand. These are the code provisions that you need to know, and this book was written to help you understand what they are and how to use them in your daily work.

Throughout my 25 years of experience in the construction industry I have seen electricians, contractors, and even engineers struggle to grasp the intention, direction, and requirements of the National Electrical Code®. Part of the problem is that people with hands-on experience in the industry usually deal with materials and installation needs from a practical point of view, not through the eyes of an electrical inspector. Contractors look for cost-efficient approaches, designers focus on aesthetics, estimators deal with unit pricing and total square footage, and electricians in the field are constantly striving to get their work done on time. The one common factor that connects all of these occupations is that any job that includes electrical work has to meet NEC® standards. It is one thing to own a code book and quite another to really understand the code and know where the various requirements for typical electrical installations can be found.

Ignorance or misinterpretation of the code can have disastrous results. Think about the cost of copper today and what would happen if you ordered the wrong cable or conduit for an installation. What if you overlooked the new code requirement for arc-fault circuit interrupters (AFCIs) for branch circuits in newly constructed dwellings and had to add them after the job was complete? Can you imagine how much time and money that would cost? Ask yourself this: Do you really understand the new section that was added in NEC® 210.4(D) that requires all the associated conductors of a multiwire branch circuit, including the grounded conductor, to be physically grouped together at least once by wire ties or a similar means within the branch-circuit overcurrent device enclosure? Can you quickly explain the definitions of a "neutral conductor" and a "neutral point" that were added to the 2008 NEC® Article 100?

The book you are holding right now is designed for anyone who has ever felt overwhelmed by the complexity of the National Electrical Code® or become frustrated trying to locate a specific section of the code. The purpose is to protect you from costly violations and inspection rejections that can happen because you just weren't clear on the meaning behind a standard or what the current edition of the code requires.

Each chapter explains key code requirements in plain English and provides you with code updates from the 2008 edition of the NEC®. There are Trade Tips throughout the book that provide hints about applying the code and "Did You Know?" callout boxes with quick references and Do's and Don'ts for the electrical trade. You'll have the latest requirements at your fingertips for general installation branch circuits, feeder, and service calculations, overcurrent protection, grounding, and even generators and transformers. And don't forget that the whole purpose of the NEC® is to safeguard people and property from the many hazards associated with electricity. To achieve this goal, you have to have a clear and comprehensive understanding of the NEC® as well as a working knowledge of the electrical trade. That's why this book also includes extra information on subjects such as NFPA 70E®, electrical terms, conversion and calculation tables, and a cross-reference of associated code articles.

NEC® 2008 Need To Know brings together all the elements you need to design, estimate, and install electrical systems and equipment while explaining code requirements in a way that you can easily understand so you can comply with the National Electrical Code®. It's like having your own personal condensed code book that gives you just what you need to know to optimize your time and increase your effectiveness.

I believe you are going to find this an essential asset to your daily work in the electrical industry.

Best regards to all the "Sparkies" out there!

—K. J. Keller

Acknowledgments

I would like to thank the National Fire Protection Association® for permission to reference the 2008 NEC® standards and for allowing me to reprint materials from the 2008 National Electrical Code® book for the benefit of my readers.

I also with to thank my publisher, McGraw-Hill, for the opportunity to shed light on the grayer areas of the NEC® that affect so many electrical contractors. Additionally, I appreciate the efforts of LoneWolf Enterprises and Roger Woodson in editing and preparing this finished work.

NEC® 2008 NEED TO KNOW

1

General Requirements for Electrical Installations

WHAT YOU NEED TO KNOW

As an electrician, you are responsible for the health and safety of both yourself and the public. Your job is much bigger than just running wire or hanging light fixtures. An installation or material mistake or a code violation could result in electrical shock, blast, fire, injury, or even death. For this reason, the National Fire Protection Association®, Inc., known by the acronym NFPA®, develops and publishes the National Electrical Code® (NEC®) which is a compilation of codes and standards under the auspices of the International Electrical Code. Each section of the **NEC** falls under NFPA 70® and has been developed through a committee process to establish standards for electrical installations that reduce the risk of hazards. The **National Electrical Code** is organized by articles that categorize general areas of electrical work, and each article is subdivided into sections that detail standards for specific aspects of that work.

Article 90: Introduction to the NEC

Article 90 of the **NEC** is a basic introduction to the intentions of the codes. Essentially, it describes the purpose of the **National Electrical Code** as providing a uniform and practical means of safeguarding people and equipment from electrical hazards. These safety guidelines are not meant to provide the most convenient or efficient installations and don't even guarantee good service or allow for future expansion. They are designed to provide a standard for safety that protects against electrical shock and thermal effects as well as dangerous overcurrents, fault currents, and overvoltage. The **NEC** parallels the principles for safety covered in Section 131 of the International Electrotechnical Commission Standard for electrical installations for buildings.

Article 90 provides for special permission from the authority having jurisdiction in a locality to grant an exception for the installation of conductors and equipment that are outside a building or that terminate immediately inside a building wall. The Authority Having Jurisdiction (AHJ) can also waive certain requirements of the code or allow alternative methods as long as these exceptions ensure effective safety. Administrative polices and procedures are what make the electrical code effective. Without the proper procedures and administration, the electrical code would be little more than an organized outline for good electrical procedures.

To be effective, the code must be enforced. To be fair, the rules for the administration of the code must be clear to everyone who works with it. Administrative policies dictate the procedure for code enforcement, interpretation, and implementation. Essentially, then, the code is intended to provide a standard that government bodies can enforce. Some of the **NEC** rules are mandatory and can be identified by the terms *shall* or *shall not*.

Fast Fact

Local code-enforcement officers are permitted to interpret the **National Electrical Code**. While the interpretation of the code officer may be quite different than your understanding of a code, it is the code officer's option to determine the meaning of the code and your responsibility to comply with that interpretation.

Other rules describe actions that are allowed but are not mandatory, and these are characterized by the terms *shall be permitted or shall not be required*. Throughout the code book, principles are explained or cross-referenced to related parts of the code in the form of Fine Print Notes (FPN); these informational notes are not enforceable.

Also found in Article 90 are general explanations of wiring planning. For example, metric units of measure (SI) are listed first, with inch-pound units following, and trade practices that are used in trade sizing.

Article 100: Definitions

Chapter 1 of the **NEC** begins with Article 100, which is an alphabetical listing of definitions that are key to the proper application of the code. Some of these definitions are simple enough: for example, "dwelling, one-family" is described as a "building that consists solely of one dwelling unit." Many of the descriptions are written in more complex terms. An "outlet," for example, is defined as "a point on the wiring system at which current is taken to supply utilization equipment," and a "power outlet" is "an enclosed assembly that may include receptacles, circuit breakers, fuse holders, fused switches, buses, and watt-hour meter mounting means; intended to supply and control power to mobile homes, recreational vehicles, park trailers, or boats or to serve as a means for distributing power required to operate mobile or temporarily installed equipment."

As a licensed electrician, it is assumed that you have a clear understanding of electrical materials, terminology, and applications as they are defined in the **NEC**. As a case in point, you may often refer to conductors that are listed or rated for a certain use as "approved" materials. Yet, the **NEC** definition of "approved" does not pertain to rated or tested materials; rather, it means "acceptable to the Authority Having Jurisdiction." One the other hand, sometimes practices that are common in the field are not specifically required or defined in the **NEC**. For example, a bonding jumper is described in the **NEC** as a reliable conductor that is used to ensure the required electrical conductivity between metal parts that are electrically connected. Tradespeople traditionally use a green insulated wire or a black wire marked with green tape, but nowhere in the definition is a specific color required for a bonding jumper.

The 2008 edition of the NEC contains a number of changes in definitions. Some of these changes, such as the terminology relating to "grounding" and "bonding," which are first listed in Article 100, were revised throughout the code. Definition revisions are listed for you below, along with an analysis of many of these changes. Alterations in the codes are underlined to make it easier for you to identify definition changes.

> ***2005 NEC**: Article 100 Bonding (Bonded). The permanent joining of metallic parts to form an electrically conductive path that ensures electrical continuity and the capacity to conduct safely any current likely to be imposed.

> **2008 *NEC***: **Article 100: Bonded** (Bonding). Connected to establish electrical continuity and conductivity.

This revision was written to apply in a general manner throughout the NEC; it describes the purpose and function of bonding. You should note that there are specific bonding requirements to minimize the difference of potential (voltage) between conductive components.

> **2008 *NEC***: **Article 100: Bundled.** Cables or conductors that are physically tied, wrapped, taped, or otherwise periodically bound together.

This term definition has be relocated from Article 520.2 to Article 100.

CODE UPDATE

New definition added to Article 110:

Branch-Circuit Overcurrent Device. A device capable of providing protection for service, feeder, and branch circuits and equipment over the full range of overcurrents between its rated current and interrupting rating. Branch-circuit overcurrent protective devices are provided with interrupting ratings appropriate for the intended use but no less than 5,000 amperes.

This new definition was added to clarify that a branch-circuit overcurrent device is capable of providing protection for service, feeder, and branch circuits.

CODE UPDATE

These new definitions have been added to Article 110:

Clothes Closet. A non-habitable room or space intended primarily for storage of garments and apparel.

This new definition was added to clearly distinguish a clothes closet from other types of closets.

Electrical Power Production and Distribution Network. Power production, distribution, and utilization equipment and facilities, such as an electric utility system that delivers electric power to the connected loads, that is external to and not controlled by an Interactive System.

This new definition correlates with the definition in section 2.41 of Underwriter Laboratory Standard 1741: Inverters, Converters and Controllers for Use in Independent Power Systems.

Equipotential Plane. An area where mesh or other conductive elements embedded in or placed under a concrete or other conductive surface are bonded together, bonded to all metal structures and fixed nonelectrical equipment that may become energized, and connected to the electrical grounding system.

This new definition combines two concepts into one common term.

---

*2005 **NEC**: Article 100: Dwelling Unit. A single unit, providing complete and independent living facilities for one or more persons, including permanent provisions for living, sleeping, cooking, and sanitation.

2008 *NEC*: Article 100: Dwelling Unit. One or more rooms arranged for complete, independent housekeeping purposes, with space for eating, living, and sleeping; facilities for cooking; and provisions for sanitation.

This revision simplifies what a dwelling unit is and meets the needs of the **NEC**, NFPA 1, NFPA 101, and NFPA 5000.

*2005 **NEC**: Article 100: Equipment. A general term including material, fittings, devices, appliances, luminaires (fixtures), apparatus, and the like used as a part of, or in connection with, an electrical installation.

2008 *NEC*: **Article 100: Equipment.** A general term including material, fittings, devices, appliances, luminaires (fixtures), apparatus, machinery, and the like used as a part of, or in connection with, an electrical installation.

The word "machinery" is now included in this definition to ensure that electrical machinery is included along with electrical equipment. For example, Section 110.2 requires conductors and electrical equipment required by the **NEC** to be acceptable only if approved. Adding the term "machinery" now clarifies that industrial equipment installations are covered by the **NEC**.

*2005 **NEC**: Article 100: Ground. A conducting connection, whether intentional or accidental, between an electrical circuit or equipment and the earth or to some conducting body that serves in place of the earth.

2008 *NEC*: **Article 100: Ground.** The earth.

*2005 **NEC**: Article 100: Grounding Electrode. A device that establishes an electrical connection to the earth.

2008 *NEC*: **Article 100: Grounding Electrode.** A conducting object through which a direct connection to earth is established.

This definition has been rewritten to better describe the function of a grounding electrode.

*2005 **NEC**: Article 100: Grounding Electrode Conductor. The conductor used to connect the grounding electrode(s) to the equipment grounding conductor, to the grounded conductor, or to both, at the service, at each building or structure where supplied by a feeder(s) or branch circuit(s), or at the source of a separately derived system.

2008 *NEC*: **Article 100: Grounding Electrode Conductor.** The conductor used to connect the grounding electrode(s) to a system conductor or to equipment.

*2005 **NEC**: Article 100: Grounding Conductor, Equipment. The conductor used to connect the non-current-carrying metal parts of equipment, raceways, and other enclosures and to the system grounded conductor or to grounding electrode conductor, or both, at the service equipment or at the source of a separately derived system.

2008 *NEC*: **Article 100: Grounding Conductor, Equipment.** The conductive path installed to connect normally non-current-carrying metal parts of equipment together and to the system grounded conductor or to grounding electrode conductor. FPN No. 1: It is recognized that the equipment grounding conductor also performs bonding. FPN No.2: See 250.8 for a list of acceptable equipment grounding conductors.

*2005 **NEC**: Article 100: Handhole Enclosure. An enclosure identified for use in underground systems, provided with an open or closed bottom, and sized to allow personnel to reach into, but not enter, for the purpose of installing, operating, or maintaining equipment or wiring or both.

2008 *NEC*: **Article 100: Handhole Enclosure.** An enclosure for use in underground systems, provided with an open or closed bottom, and sized to allow personnel to reach into, but not enter, for the purpose of installing, operating, or maintaining equipment or wiring or both.

The requirement of identification has been removed from definition.

CODE UPDATE

These new definitions are now included in the 2008 **NEC**, Article 100:

Intersystem Bonding Termination. A device that provides a means of bonding communications equipment at the service equipment or the disconnecting means for any additional buildings or structures.

This new definition describes a device that connects communications equipment to the service equipment or the disconnecting means for additional buildings or structures in accordance with Part II in Article 225.

Kitchen. An area with a sink and permanent facilities for food preparation and cooking.

*2005 **NEC**: Article 100: Luminaire. A complete lighting unit consisting of a lamp or lamps, together with the parts designed to distribute the light to position and protect the lamps and ballast (where applicable) to connect the lamps to the power supply.

2008 *NEC*: **Article 100: Luminaire.** A complete lighting unit consisting of a light source, such as a lamp or lamps, together with the parts designed to position the light source and connect it to the power supply. It may also include parts to protect the light source, ballast, or distribute the light. A lampholder itself is not a luminaire.

The idea behind this wording change is to clarify that a lamp holder is *not* a luminaire.

*2005 **NEC**: 550.2: Manufactured Home. A structure, transportable in one or more sections, that is 2.5 meters (8 body feet) or more in width or 12 meters (40 body feet) or more in length in the traveling mode or, when erected on site, is 30 meters^2 (320 feet^2) or more; which is built on a chassis and designed to be used as a dwelling, with or without a permanent foundation, when connected to the required utilities, including the plumbing, heating, air-conditioning, and electrical systems contained therein. Calculations used to determine the number of square meters (square feet) in a structure will be based on the structure's exterior dimensions, measured at the largest horizontal projections when erected on site. These dimensions include all expandable rooms, cabinets, and other projections containing interior space, but do not include inside bay windows.

2008 *NEC*: **Article 100: Manufactured Home.** A structure, transportable in one or more sections, that, in the traveling mode, is 8 body feet (2.4 meters) or more in width or 40 body feet (12.2 meters) or more in length or, when erected on site, is 320 feet^2 (29.7 meters^2) or more and that is built on a permanent chassis and designed to be used as a dwelling, with or without a permanent foundation, when connected therein. The term "manufactured home" includes any structure that meets all the provisions of this paragraph except the size requirements and with respect to which the manufacturer voluntarily files a certification required by the regulatory agency, and except that such term shall not include any self-propelled recreational vehicle. Calculations used to determine the number of square feet (square meters) in a structure are based on the structure's exterior dimensions, measured at the largest horizontal projections when erected on site. These dimensions include all expandable rooms, cabinets, and other projections containing interior space, but do not include bay windows.

This change is intended to provide consistent meaning of defined terms throughout the National Fire Codes.

> *2005 **NEC**: Article 100: Qualified Person. One who has skills and knowledge related to the construction and operation of the electrical equipment and installations and has received safety training on the hazards involved.

> 2008 *NEC*:Article 100: **Qualified Person.** One who has skills and knowledge related to the construction and operation of the electrical equipment and installations and has received safety training <u>to recognize and avoid</u> the hazards involved.

CODE UPDATE

New definitions added to Article 100 include:

Short-Circuit Current Rating. The prospective symmetrical fault current at a nominal voltage to which an apparatus or system is able to be connected without sustaining damage exceeding defined acceptance criteria.

    This provides a definition for the phrase used in 110.10 and in other areas in the code.

Surge Protective Device (SPD). A protective device that is used for limiting transient voltages by diverting or limiting surge current; it also prevents continued flow of follow current while remaining capable of repeating these functions and designated as follows:

    Type 1:Permanently connected SPDs intended for installation between the secondary of the service transformer and the line side of the service disconnect overcurrent device.

    Type 2 : Permanently connected SPDs intended for installation on the load side of the service disconnect overcurrent device, including SPDs located at the branch panel.

    Type 3: Point of utilization SPDs.

    Type 4: Component SPDs, including discrete components, as well as assemblies.

    FPN No. 1: For further information on Type 1, Type 2, Type 3, and Type 4 SPDs, see UL 1449, Standard for Surge Protective Devices.

    This new definition now replaces the term "transient voltage surge suppressor" used in Article 285.2.

(continued on next page)

Ungrounded. Not connected to ground or a conductive body that extends the ground connection.

The term "ungrounded" is used extensively in the **NEC**, but it was not defined in early editions.

Utility-Interactive Inverter. An inverter intended for use in parallel with an electric utility to supply common loads and sometimes deliver power to the utility.

The reason that this definition was added is because the term is used in revised Articles 690 and 692 and Article 705 with respect to the interconnection of systems and equipment for use with distributed energy resources. This definition also correlates with the definition in Section 2.4 of Underwriter Laboratory Standard 74, *Inverters, Converters and Controllers for Use in Independent Power Systems.*

Article 110: General Installation Requirements

One of the key elements of Article 110 is that it establishes uniform, approvable standards for materials and installations. For example, Article 110.3 lists the considerations for material installations, such as the suitability of the installation, the mechanical strength and durability of materials, the degree of electrical insulation provided, possible arcing effects of the materials or installation, and what constitutes suitable wiring methods. Basically, this article of the **NEC** spells out the fundamentals for electrical work.

Here is an easy example. Article 110.12 requires that electrical systems must be installed in a neat and workmanlike manner. Does this mean conduits must be level or that conductors must be wire-tied together? No. It means that unused cable or raceway openings in equipment cases, boxes, or auxiliary gutters must be effectively closed to provide the equivalent protection of the wall of the equipment. It also means that conductors in

Fast Fact

None of the requirements listed in Article 110 apply to communications circuits. Those standards are listed separately in Chapter 8.

CODE UPDATE

These two new definitions were added to Article 110:

Neutral Conductor. The conductor connected to the neutral point of a system that is intended to carry current under normal conditions.

Neutral Point. The common point on a wye-connection in a polyphase system or midpoint on a single-phase, 3-wire system, midpoint of a single-phase portion of a 3-phase delta system, or a midpoint of a 3-wire, direct current system.

FPN: At the neutral point of the system, the vectorial sum of the nominal voltages from all other phases within the system that utilize the neutral, with respect to the neutral point, is zero potential.

Adding these two new definitions identifies what constitutes a neutral conductor and a neutral point.

---

underground or subsurface enclosures that people must enter to install or maintain equipment have to be racked. This provides safe and easy access. Also, internal parts of electrical equipment, such as busbars, insulators, or wiring terminals, cannot be damaged, bent, broken or cut, or contaminated by materials such as paint or plaster, abrasives, or corrosive residues. This is a commonsense requirement, because this kind of damage could affect the safe operation or mechanical strength of the equipment.

Arc Flash and Arc Blast

Article 110.16 requires that switchboards, panelboards, and motor or industrial control panels, in any installation other than a dwelling unit, must be labeled with a warning indicating that a potential arc flash hazard exists.

Revision to Article 110.16–Flash Protection

Electrical equipment such as... shall be field marked to warn qualified persons of potential electric arc flash hazards.

Two changes were made to 110.16 to provide appropriate and consistent application to field installations that qualify for the warning labels required by this section. The first revision broadens the requirement by including all types of equipment that would qualify for the field-applied arc-flash warning labels. Previously the requirements of this section were limited to equipment that was actually identified in the rule. By including the words "**equipment such as**" the concept is expanded to all equipment types that are likely to require examination, adjustment, servicing, or maintenance while energized. This requirement applies to equipment such as enclosed circuit breakers, some types of transformers, and other equipment that was not specifically included in the previous text. The second revision puts a limitation on the types of dwelling occupancies in which these labels for equipment are required. For example, multiple-occupancy dwelling structures such as apartment buildings, where the service and other equipment can be large, require arc-flash warning labels. However, arc-flash warning-label requirements do not apply to one- and two-family dwelling units. This specific rule only applies to installations where field-applied arc-flash warning labels are required.

The **National Fire Protection Association** has established a standard for electrical safety in the workplace, NFPA 70E®, which outlines the procedures required to determine safe working distances and the appropriate personal protective equipment (PPE) to be used if electrically live components cannot be placed in an electrically safe work condition. **NFPA 70E** is used to expand and clarify the generalized installation requirements listed in the **NEC**. In 2002, the **NEC** referenced **NFPA 70E** for the

FIGURE 1.1

Sample arc-flash warning label.

first time. In the 2005 **NEC** Section 110.16, Fine Print Note No. 1 refers the reader to **NFPA 70E** for assistance in determining severity of potential exposure, planning safe work practices, and selecting personal protective equipment. **NFPA 70E** spells out flash hazardous conditions and provides the calculations necessary to establish safe approach boundaries and adequate PPE for working on energized equipment. The definitions of these conditions are as follows:

Flash Hazard. A dangerous condition associated with the potential for the rapid release of energy caused by an electrical arc that suddenly and violently changes material(s) into a vapor.

Flash Protection Boundary. This boundary is the closest anyone can approach equipment that poses a flash hazard without the use of PPE.

Limited Shock Approach Boundary. This boundary can only be crossed by a qualified electrical worker or a person who is accompanied by a "qualified" worker.

Restricted Shock Aproach Boundary. This boundary may only be crossed by an "electrically qualified" person who is using adequate PPE, shock-prevention equipment, and techniques.

All of these boundaries should be posted on the related equipment using arc flash labels.

You may be asking yourself; what does **NFPA 70E** have to do with the NEC? After all, you know that FPNs in the **National Electrical Code** are not mandatory. Bear in mind that when you install electrical systems, not only do you have to comply with **NEC** requirements, but you have to work in a manner that meets OSHA standards of safety. It was OSHA that implored NFPA to develop a standard to address safe electrical work practices. Remember, NFPA issues the **NEC**, and it took this issue so seriously that it developed **NFPA 70E** as a means of connecting the electrical code with health safety standards.

Trade Tip

There is no required standard for arc-flash label designs.

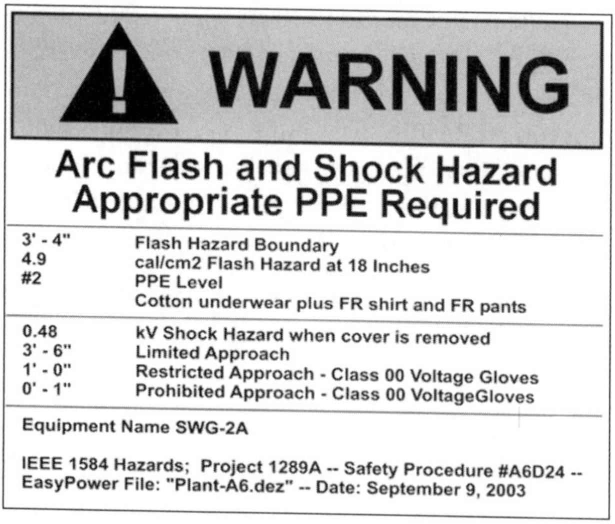

FIGURE 1.2

Arc-Flash Hazard Label. A comprehensive label lists the flash hazard boundary, limited approach distance, and the class of gloves and PPE required.

Although **NFPA 70E** is not incorporated in OSHA 29 CFR 1910.6, it does detail and clarify such enforceable OSHA standards as:

29 CFR 1910.132 (d)(1): requires employers to perform a personal protective equipment (PPE) hazard assessment to determine necessary PPE.

29 CFR 1910.269 (l)(6)(iii): requires employers to ensure that each employee working at electric-power generation, transmission, and distribution facilities who is exposed to the hazards of flames or electric arcs does not wear clothing that could increase the extent of injury when exposed to such a hazard.

29 CFR 1910.335 (a)(1)(i): states that employees working in areas where there are potential electrical hazards shall use electrical protective equipment appropriate for the specific parts of the body for the work being performed.

29 CFR 1910.335 (a)(1)(iv): requires employees to wear nonconductive head protection whenever exposed to electric shock or burns due to contact with exposed energized parts.

Did You Know?

OSHA is currently in the process of training its compliance inspectors to use **NFPA 70E** as the benchmark to gauge "safe" working conditions for equipment with the potential for arc flash or arc blast.

29 CFR 1910.335 (a)(1)(v): states that employees shall wear protective equipment for the eyes or face wherever there is danger of injury to the eyes or face from electric arcs or flashes or from flying objects resulting from an electrical explosion

So there you have it. One little article with some basic language about flash protection and a Fine Print Note serve as a conduit connecting **NFPA 70E** with OSHA-enforceable standards.

New Article 110.20

There is a new article that has been added to the **NEC**, Article 110.20, which covers enclosure types: Enclosures shall be marked with an Enclosure Type number as shown in Table 110.20. This new article relocates an existing chart, Table 430.91, to Article 110 because the table includes enclosure-type designations and information about their use. The table is now identified as Table 110.20.

Electrical enclosures are required to be suitable for the environment where they are installed. This change now provides consistency between the UL General Information for Electrical Equipment Directory (White Book) category AALZ and the **NEC**, which have general applications to all electrical equipment, not just equipment associated with the types of motor installations covered by Article 430. More clarity on this subject was

Did You Know?

All of the equipment types in Table 110.20 are required to use a Type number marking by existing industry product standards.

FIGURE 1.3

Table 430.91 is now NEC Table 110.20

NEC © Table 110.20 Enclosure Selection

Provides a Degree of Protection Against the Following Environmental Conditions	For Outdoor Use									
	Enclosure-Type Number									
	3	3R	3S	3X	3RX	3SX	4	4X	6	6P
Incidental contact with the enclosed equipment	X	X	X	X	X	X	X	X	X	X
Rain, snow, and sleet	X	X	X	X	X	X	X	X	X	X
Sleet*	—	—	X	—	—	X	—	—	—	—
Windblown dust	X	—	X	X	—	X	X	X	X	X
Hosedown	—	—	—	—	—	—	X	X	X	X
Corrosive agents	—	—	—	X	X	X	—	X	—	X
Temporary submersion	—	—	—	—	—	—	—	—	X	X
Prolonged submersion	—	—	—	—	—	—	—	—	—	X

For Indoor Use

Provides a Degree of Protection Against the Following Environmental Conditions	Enclosure Type Number									
	1	2	4	4X	5	6	6P	12	12K	13
Incidental contact with the enclosed equipment	X	X	X	X	X	X	X	X	X	X
Falling dirt	X	X	X	X	X	X	X	X	X	X
Falling liquids and light splashing	—	X	X	X	X	X	X	X	X	X
Circulating dust, lint, fibers, and flyings	—	—	X	X	—	X	X	X	X	X
Settling airborne dust, lint, fibers, and flyings	—	—	X	X	X	X	X	X	X	X
Hosedown and splashing water	—	—	X	X	—	X	X	—	—	—
Oil and coolant seepage	—	—	—	—	—	—	—	X	X	X
Oil or coolant spraying and splashing	—	—	—	X	—	—	—	—	—	X
Corrosive agents	—	—	—	X	—	—	X	—	—	—
Temporary submersion	—	—	—	—	—	X	X	—	—	—
Prolonged submersion	—	—	—	—	—	—	X	—	—	—

*Mechanism shall be operable when ice covered.

FPN: The term *raintight* is typically used in conjunction with Enclosure Types 3, 3S, 3SX, 3X, 4, 4X, 6, and 6P. The term *rainproof* is typically used in conjunction with Enclosure Types 3R, and 3RX. The term *watertight* is typically used in conjunction with Enclosure Types 4, 4X, 6, 6P. The term *driptight* is typically used in conjunction with Enclosure Types 2, 5, 12, 12K, and 13. The term *dusttight* is typically used in conjunction with Enclosure Types 3, 3S, 3SX, 3X, 5, 12, 12K, and 13.

provided by relocating the requirements of 430.91 and Table 430.91 into the general application sections of Article 110 and spelling out the types of equipment that are covered.

Article 110.26: Working Spaces

Arc-flash and arc-blast protection emphasize the importance of adequate working space around enclosed equipment. It's not about how much easier it is to work with a little space around you or about not being so cramped up that you risk contact with live parts that can do much more than just give you a little zap. As an electrician you may either install or service electrical equipment that has the potential to cause serious injury or death. Let's examine how the arc-protection requirement ties into ample work-space requirements by taking a quick look at the effects of electricity on the human body.

As an electrician, you may have taken a little "hit" on at least one occasion and walked away just fine. When does a zap cross over into a severe injury? Do you think it's at 220 volts or 400 volts? The fact of the matter is that the amount of internal electrical current that your body can tolerate and still enable you to control the muscles of your arms and hands is some-where around 10 milliamperes (mA). Remember your basic electrical math? A milliampere equals 1/10,000 of an amp. Currents above 10 mA can paralyze or "freeze" your muscles so that you can't release a tool or wire. As a matter of fact, these muscles will contract in response to the current so you end up clenching the electrified object even more tightly. This results in extended exposure to the shocking current. If you can't break contact with the energized parts, the current will continue racing through your body for a longer time. Now you have respiratory paralysis kicking in, because the muscles in your diaphragm that control your breathing can't move and you stop breathing for a period of time. People have stopped breathing after being shocked with currents from voltages as low as 49 volts.

Did You Know?

1/10 of an ampere of electricity going through the body for just 2 seconds is enough to cause death.

TABLE 1.1 *Work Space Minimum Clear Distances*

Nominal Voltage to Ground	Condition 1	Condition 2	Condition 3
0–150 volts	3 feet	3 feet	3 feet
151–600 volts	3 feet	3 1/2 feet	4 feet

- Condition 1: represents electrical equipment that is mounted or set on one wall and the wall on the opposite side is insulated (ungrounded parts).
- Condition 2: pertains to electrical equipment that is mounted or set on one wall and the wall on the opposite side is grounded.
- Condition 3: is a condition where electrical equipment is mounted or set on one wall and additional electrical equipment is mounted or set on the opposite side of the room.

Now you can see the importance of complying with the clearance requirements in Article 110-26 of the **NEC**. Let's look at these three clearances in detail, beginning with depth of working space in Section 110.26(A)(1). The depth of the working space in front of exposed live parts must be at least as great as the distances outlined in Table 110-26(A)(1).

As you can see, there are three conditions listed that cover two voltage levels of 150V or less and 151V to 600V to ground. The everyday way to apply these conditions is that the clearances of a 120V/208V, 3-phase, 4-wire system fall under the 150V-to-ground or less category, or 3 feet. A 277V/480V, 3-phase, 4-wire system would be categorized in the 151V to 600V-to-ground classification.

To comply with the depth requirements, measure the distance from any live parts or from the front of the enclosure that encloses the live parts to the opposing wall. These conditions are *not* Fine Print Notes, which means that they are mandatory considerations.

Condition 1 describes a scenario where the electrical equipment is installed in or on a wall that faces an insulated wall that is made of wood or metal studs and sheetrock or wood panels. If you make contact with the insulated wall while touching any live parts of the equipment, you're isolated from the grounded slab or earth, so Condition 1 allows for a reduced working space.

Condition 2 involves a situation in which the electrical equipment is installed on a wall that faces a conductive or grounded wall. To be considered conductive, a wall must be made of concrete, brick, or tile, which means that it will connect the body to ground if it is touched. So, if you make contact with this type of wall while you are touching a live part or conductor, you will create a circuit path to ground, which could lead to electrocution. This potentially dangerous situation prompts the need for a larger work space if the live equipment voltages range from 151V to 600V.

Condition 3 pertains to electrical equipment that is installed in or on a wall that faces another wall of electrical equipment so that there are live parts on both sides of the room. In this case, you could be subjected to phase-to-phase voltage or phase-to-ground voltage sufficient to cause a fatal shock while working on energized equipment. Under these conditions, the **NEC** requires a greater safe clearance.

There are three exceptions to the depth requirements:

- Dead-front assemblies where the connections are accessible from locations other than just the front or sides
- Low-voltage energized parts less than 30 volts rms, 42 volts peak, or 60 volts dc
- Existing building conditions where equipment is being replaced

The next dimension of safe working space, width, is listed in Section 110-26(A)(2). This standard requires that you have a measured clearance from the sides of live parts to the wall that is either the same as the width of the equipment or 30 inches, which ever is greater. Additionally, you have to allow enough clearance so that equipment doors and hinged panels in the workspace can open to at least 90 degrees. The reason for this is to ensure that you have enough room to work on overcurrent-protection devices without putting your body between the panel door and the panelboard.

The final requirement, Section 110-26(A)(3), involves clearance height. As a general rule, you have to maintain a minimum headroom clearance of 6 feet from the floor up to the ceiling or to any overhead obstruction. This requirement applies to service equipment, switchboards, panelboards, and motor control centers and is designed to protect you from accidentally touching grounded objects and exposed live parts at the same time. The bottom line is that electricians or maintenance workers should never have to stoop or bend down to gain access to service or repair components inside electrical equipment.

PUTTING IT ALL TOGETHER

The purpose of the **NEC** is to provide practical safeguarding from electricity to both people and property. Even though the standards may not always outline the most cost-effective, convenient or even efficient prqctices, they do create a consistent guideline for safe use and exposure to electricity. As an electrician, it is your responsibility to not only know and understand the codes but to practice and utilize them as well. The easiest way to do this is to understand the intention behind the codes and how the standards apply to your planning, installations, and workmanship.

CHAPTER

2

Branch Circuits and Feeders

It is important to understand the difference between branch circuits and feeders and the importance of sizing, loads, and location as they apply to these circuits. Branch circuits are defined in Article 100 as the circuit conductors between outlets and the final overcurrent device that protects the circuit. Feeders are comprised of all the circuit conductors between service equipment, the source of a separately derived system, or another power supply source. They are the final branch-circuit overcurrent protection device. The bottom line is that a feeder occupies the spaces *between* a power-distribution system.

TERMS TO KNOW

Following are some of the key terms you need to be familiar with:

Branch Circuit: The section of a wiring circuit between the final set of fuses or final breaker and the outlets it supplies.

Dwelling Unit: One or more rooms arranged for complete, independent housekeeping purposes, with space for eating, living, and sleeping; facilities for cooking, and provisions for sanitation.

GFCI: Ground-fault circuit interrupter, a unique type of electrical receptacle or outlet that can stop electrical power within milliseconds as a safety precaution.

Grounded Conductors: Conductors that are connected to the earth or to a conductive body that extends the ground connection.

Neutral Conductor: Any current-carrying conductor that has a net voltage to ground of zero measured at the service ground bus.

Overcurrent: A condition that exists on an electrical circuit when the normal load current is exceeded. Overcurrents take on two seperate characteristics, overloads and short circuits

BRANCH CIRCUITS

Article 210 contains standards for conductor sizing and identification, overcurrent protection, and GFCI protection for branch circuits, as well as requirements for receptacle outlets and lighting outlets. Various methods of identifying grounded conductors are determined based on the size of the grounded conductor and the type of cable or cord. Additionally, the terminals of grounded conductors have their own identification requirements.

Branch circuits are rated based on the maximum amperage rating allowed for the overcurrent device. With the exception of individual branch circuits, the ratings allowed are 15, 20, 30, 40, and 50 amps. If for some reason higher amperage conductors are used, the amperage rating or setting of the overcurrent device will determine the circuit rating. The exception is multi-outlet branch circuits that are greater than 50 amps; they are allowed to supply non-lighting outlet loads in industrial applications where only qualified people will be servicing the equipment.

Table 210.2 identifies the various types of specific-purpose branch circuits and references where in the **NEC** these types of installation standards are found. For example, air-conditioning and refrigeration equipment requirements are located in **NEC** Sections 440.6, 440.31, and 440.32, while motor circuits and controller regulations are outlined in Article 430.

NEC 210.4 Multiwire Branch Circuits

A multiwire branch circuit is described as having two or more ungrounded conductors with voltage between them and a grounded conductor that has equal voltage between it and each of the ungrounded conductors and is connected to a neutral or grounded conductor in the same system. The conductors can only supply line-to-neutral loads and must originate from the same panel. Multiwire branch circuits are beneficial because they result in a lower circuit voltage drop and therefore result in cost savings.

One of the most common and most important types of multiwire branch circuit is the Ground Fault Circuit-Interrupter or GFCI. As an electrician, you probably already know that a GFCI functions to de-energize a circuit or a part of a circuit if the current to ground exceeds the standards for a Class A device. But do you know what that value is? For a Class A device, a GFCI will trip if the current to ground is between 4mA and 6mA. You should think of this type of branch circuit as a safety device and know that

---

CODE UPDATE

In the 2008 edition of **NEC** 210.4 the standard for "simultaneous disconnecting means" for multiwire branch circuits has been expanded to include the requirement that all multiwire branch circuits have a means to simultaneously disconnect all ungrounded conductors. This must take place at the point where the branch circuit originates. This process of disconnecting each ungrounded conductor of a multiwire branch circuit can be achieved by use of common trip 2-pole or 3-pole circuit breakers or single-pole circuit breakers that are identified with a handle tie. Based on **NEC** 240.15(B)(1), you can use individual single-pole circuit breakers with handle ties identified for this purpose or a breaker with a common internal trip. This section of the **NEC** also now requires that all of the conductors associated with a particular multiwire branch circuit be physically grouped together with wire ties or a similar means at the point of origin. An exception is made for circuit conductors that are contained in a single raceway or cable, because this makes the grouping obvious.

Article 210.8 outlines eight specific locations where this type of circuit protection is required. Any dwelling unit with 125-volt, single-phase, 15- or 20-amp receptacles installed in the following locations requires a GFCI:

- Bathrooms
- Garages or "accessory" buildings that are not intended to be lived in and that have a floor that is located at or below grade
- Outdoor locations, with the exception of receptacles that are not readily accessible and that are supplied by a dedicated branch circuit for electric snow-melting or deicing equipment
- Crawl spaces that are at or below grade level
- Unfinished basements (see exceptions below)
- Kitchen-countertop receptacles
- Laundry and utility rooms and wet-bar sink areas with receptacles installed within 6 feet of the outside edge of the sink
- Boat houses and boat hoists

In the old edition of the **NEC**, receptacles that were not "readily accessible" were excluded from the regulations regarding ground-fault protection. This terminology was deleted because the interpretation of "readily accessible" led to inconsistencies in how the requirements were applied. For example, to a child a garage-door opener might not be readily accessible when it would be to an adult. The section of the code is focused on shock protection, and more clarification was needed to provide consistency with the GFCI receptacle requirements in 210.8(A)(7). In that section of the code, product-safety standards for appliances in terms of this exception require appliances to be manufactured with insulation dielectric leakage levels that do not exceed 0.5 mA, which is far below the 4 to 6 mA rating of a

---

CODE UPDATE

Two exceptions to Article 210.8(A)(2) in the previous edition of the **NEC** have been deleted and additional text has been added to 210.8(A)(5), which specifies that any receptacles installed under the exceptions to 210.8(A)(5) will not be considered as meeting the requirements of 210.52(G).

CODE UPDATE

In the current 2008 version of **NEC** 210.8, revisions were made to the requirements for outdoor GFCI protection of 15-amp and 20-amp 125-volt receptacles. In the old edition, such receptacles "outdoors in public spaces" were required to have GFCI protection. The new change requires this protection in any outdoor location, unless for snow-melting or deicing equipment and industrial units. Additionally, the 2008 publication added a new subdivision to 210.8 that requires GFCI protection for all 15A and 20A, 125V receptacles that are installed within 6 feet of the outside edge of the sink, even in non-dwelling units. Exceptions are provided for industrial laboratories and patient-care areas of health-care facilities.

GFCI, so it became obvious that there was no longer a need for either of these exceptions.

Protection by ground-fault circuit interrupters is not related to the location of the receptacle. For example, if cord- and plug-connected equipment appliances have abnormal or excessive leakage current levels that will trip the GFCI, then protection needs to be provided. Consider that the maximum leakage current for a typical cord and plug appliance is 0.5 mA and the trip range for Class A GFCI devices is 4-6 mA. In order for the appliance to trip a GFCI protective device, the leakage current levels would have to be 8 to 12 times the level that is acceptable by the product standard. Obviously, this creates safety concerns. Just because a receptacle is not readily accessible does not reduce the need for shock protection for someone using the appliance. One the other hand, receptacles that supply just a permanently installed fire- or burglar-alarm system in an unfinished basement do not have to have GFCI.

Non-dwelling units with 125-volt, single-phase, 15 or 20 amp receptacles require GFCI's in following five locations:

1. Bathrooms
2. Commercial or institutional kitchens with a sink and permanent provisions for preparing and cooking food
3. Rooftop receptacles

4. Outdoor receptacles used for heating and refrigeration needs that are outlined in **NEC** Article 210.63

5. Outdoor locations in any space this is used by or is accessible to the public

All multiwire branch-circuit conductors have to originate from the same panelboard or distribution equipment in order to prevent inductive heating and reduce conductor impedance for fault currents.

The new grouping section in the 2008 **NEC** is intended to provide clearer means for workers and inspectors to identify the grounded conductor of a mutliwire branch circuit within the equipment where the circuit originates. This change will result in all associated conductors of a multiwire branch circuit, including the grounded conductor, being physically grouped together at least once within the branch-circuit overcurrent device enclosure to provide fast, easy visual recognition. Electricians will sometimes use wire ties to group conductors of several branch circuits together when they lace equipment circuits in conductor termination. Unfortunately, this habit made it difficult to tell which grounded conductors were associated with the ungrounded conductors of the same multiwire branch circuit in the equipment. Thanks to this new grouping standard, overall safety has been enhanced, making it easier for installers and inspectors to identify multiwire branch-circuit conductors within the equipment where the conductors originate.

The exception to this rule relaxes this requirement in situations where the origination of circuit conductors from a cable or raceway makes the origination obvious without additional grouping. Examples of this exception

---

CODE UPDATE

NEC Section 210.4(D) on grouping is a new section that has been added to the code and requires that the ungrounded and grounded conductors of each multiwire branch circuit be grouped by wire ties or a similar means in at least one location within the panelboard. The only exception is if circuits enter from a cable or raceway that is unique to the circuit, which makes the grouping obvious.

> **CODE UPDATE**
>
> In the 2008 edition of the **NEC**, new text has been added to Article 210.7 to clarify the requirements for simultaneously disconnecting ungrounded branch-circuit conductors. It now states that where more than one branch circuit supplies more than one receptacle on the same yoke, a means must be provided at the branch circuit panelboard to simultaneously disconnect the ungrounded (hot) circuit conductors supplying the receptacles.

would be MC or AC cable assemblies or a raceway that contains only the conductors of the multiwire branch circuit.

A multiwire branch circuit can supply line-to-line equipment, such as a range or dryer, and it can also supply both line-to-line and line-to-neutral loads if the circuit is protected by a multipole circuit breaker that opens all ungrounded conductors of the multiwire branch circuit simultaneously if there is a fault condition. This is referred to as a common internal trip. This requirement exists because if the continuity of the grounded, neutral conductor is interrupted or opened, the resulting over- or undervoltage could cause a fire or destroy electrical equipment. Individual single-pole circuit breakers with identified handle ties can be used for this application, and so can breakers with a common internal trip. The purpose is to keep people from working on energized circuits they thought were disconnected.

A means to disconnect all ungrounded conductors simultaneously must be installed for multiwire branch circuits that supply devices or equipment on the same yoke. A "yoke," which is often called a "strap," is the metal mounting structure for a device such as a switch, receptacle, or even a pilot light, and it is located at the point of origin of the branch circuit.

This requirement that all hot circuit conductors that terminate on duplex receptacles on the same yoke can be disconnected simultaneously applies regardless of the type of occupancy. It does not apply to multiwire branch circuits, since these are considered one circuit.

Dwelling-unit branch circuits can only supply loads within the dwelling unit. This standard does not allow individual dwelling units to supply common-area branch circuits in two-family or multifamily dwellings. This

---

Did You Know?

The **NEC** does not require any specific color scheme for grounded neutral conductor identification.

---

keeps common-area circuits from being disconnected by individual tenants. Additionally, it prevents the circuits from being turned off by a utility company because of nonpayment of electric bills. Along the same lines, common-area branch circuits for house lighting, central alarms, fire alarms, communications, and other similar public-safety needs can not originate from any single dwelling unit.

Neutral Conductor Identification

The grounded conductor of a branch circuit must be identified in accordance with **NEC 200.6** . If premise wiring systems contain branch circuits that are supplied from more than one voltage system, each accessible, ungrounded conductor has to be identified by system.

There are a variety of acceptable methods of identification, including color-coding, marking tape, tagging, or any other means approved by the authority having jurisdiction (AHJ). Neutral conductors must be permanently posted at each branch-circuit panelboard or distribution-equipment unit.

Trade Tip

Electricians generally use the following color system for power and lighting conductor identification:

- 120/240V single-phase: black, red, and white
- 120/208V, 3-phase: black, red, blue, and white
- 120/240V, 3-phase: black, orange, blue, and white
- 277/480V, 3-phase: brown, orange, yellow, and gray or brown, purple, yellow, and gray

---

Trade Tip

If you are using an Edison-base lampholder that is rated for 120 volts, don't put it on a 277-volt circuit.

Voltage Limitations

Dwelling units and guest rooms or guest suites in hotels and motels require branch circuits with voltages that do not exceed 120 volts nominal if the circuits are for luminaires or cord-and-plug-connected loads rated not more than 1,440 volt-amperes (VA) or less than 1/4 horsepower. **NEC** 210.6(C.) provides that in non-dwelling units, you can use 277-volt, phase-to-ground circuits to supply any of the following:

- Listed electric-discharge luminaires
- Luminaires with mogul base screw shells
- Lamp holders other than the screw-shell type
- Equipment rated at 277V

Branch-Circuit Regulations

There are three subheadings in Article 210.11 that define the number of branch circuits in any given system and explain that a load that is computed on a VA/area basis has to be evenly proportioned. The load-calculation regulations for dwelling-unit branch-circuit requirements for lighting and appliances, including motor-operated appliances, are provided further on in **NEC** 220.

NEC 210.11(A) addresses the number of allowed branch circuits. A wiring system, including the branch-circuit panelboard, has to be proportioned to serve the calculated loads and can never exceed the maximums that are specified in **NEC** 220. Loads must also be evenly proportioned as outlined in **NEC** 210.11(B) and evenly distributed among the multi-outlet branch circuits with a panelboard. Branch-circuit overcurrent protection devices and circuits only have to be installed to serve connected loads.

CODE UPDATE

Branch-circuit overcurrent protection devices are also called OCPDs, and they must have an capacity of no less than 125 percent of the continuous loads plus 100 percent of the noncontinuous loads.

NEC 210.11(C) divides the branch-circuit requirements for dwelling units into three categories. The first is the small-appliance branch-circuit regulation. which requires two or more 20-amp small-appliance branch-circuit outlets . The next is laundry branch circuits, which must have one 20-amp branch circuit that does not supply any other outlets besides a laundry receptacle. The third requirement is that there must be a 20-amp branch circuit in any bathroom that does not supply any other receptacle.

Arc-fault circuit-interrupter protection is outlined in **NEC** 210.12. You need to understand that AFCI is not the same as GFCI, even though there are combination units available. The function of an AFCI, which operates at 30mA, is to protect equipment, while a GFCI, which operates at between 4 and 6 mA, is designed to protect people. An AFCI provides protection against the effects of arc faults because it identifies the unique characteristics of arcing and will de-energize a circuit when an arc fault is detected. Any of the 15- or 20-amp, 120-volt, single-phase branch circuits in a dwelling-unit bedroom must have AFCI protection. If guest rooms or guest suites include

---

CODE UPDATE

The 2008 **NEC** requires that all 15- and 20-amp branch circuits that are installed in dwelling units have arc-fault circuit interrupter protection. This new clarification in 210.12 spells out locations that require AFCI-protection requirements for branch circuits that supply outlets in dwelling-unit family rooms, dining rooms, living rooms, parlors, libraries, dens, bedrooms, sunrooms, recreation rooms, closets, hallways, and similar areas. However, it is important to note that this 120V circuit limitation means that AFCI protection isn't required for equipment such as baseboard heaters or room air conditioners that are rated at 230 volts.

---

permanent provisions for cooking, then those branch circuits and outlets must conform to the same regulations as for dwelling units. The 2008 edition of the **NEC** includes provision 210.12, which allows the AFCI device to be located any distance from the panelboard, so long as required wiring methods are used to protect the AFCI device against physical damage.

Circuit-Rating Regulations

NEC 210.19 provides the rules for minimum amperages and sizes of conductors. One of the key points that you need to be sure you understand is that before you apply any adjustments or correction factors, a branch conductor must have an allowable amperage that is not lower than the noncontinuous load of the circuit *plus* 125 percent of the continuous load.

Article 210.21 provides a number of tables that list circuit-load and sizing standards. Table 210.21(B)(2) illustrates that the maximum load on any given circuit is 80 percent of the receptacle rating and circuit rating. You might find it easiest to read the table from right to left in order to identify the load you need to supply first, then the circuit rating you need to provide afterwards, as listed in the figure below.

Looking at the table in this way, you can see that if you need to supply a 20-amp load, you need to install at least a 30-amp receptacle on a 30-amp circuit.

The next rating table in **NEC** 210.21(B)(3) lists the various sizes of circuits connected to a receptacle. Receptacle ratings for branch circuits that supply two or more receptacles have to fall within the values shown in the table. If the receptacle rating is more than 50 amps, then the receptacle cannot be rated less than half of the branch-circuit rating.

FIGURE 2.1

Maximum cord-and-plug connected load to a receptacle.

Maximum Load In Amps	Receptacle Rating In Amps	Circuit Rating In Amps
12	15	15 or 20
16	20	20
24	30	30

Receptacle Rating In Amps		Circuit Rating In Amps
15	⇨	15 or 20
20	⇨	20
30	⇨	30

FIGURE 2.2

Receptacle ratings for associated circuit sizes.

The total rating of utilization equipment fastened in place, with the exception of light fixtures, can not exceed 50 percent of the branch-circuit amperage rating when lighting units, cord-and-plug-connected equipment that isn't fastened in place, or both of these are also supplied. The reason behind this is to prevent a circuit overload when an additional load is added—for example, if someone plugs in a vacuum cleaner. The solution is to separate circuits by putting lights on one circuit, dedicated loads (fastened in place) on different circuits, and convenience receptacles on separate circuits.

Cord-and-plug-connected equipment that is not fastened in place, such as a table saw, for example, cannot have an amperage rating that is more than 80 percent of the branch-circuit rating. Portable equipment, such as a hair dryer, may have a UL listing up to 100 percent of the circuit rating, but you have to remember that the **NEC** is an installation standard, not a product standard. As an electrician, you have no way of knowing if a circuit will be used at 80 percent of the branch-circuit rating, but you must install to that standard.

Fast Fact

Everything you need to know on a regular basis about branch-circuit requirements is summarized in **NEC** Table 210.24. Simply look for the circuit rating, which is based on the load you need to supply, and the table indicates the minimum conductor and tap sizes, overcurrent protection, and maximum load. It also lists which lampholders are permitted and what the receptacle rating must be.

FIGURE 2.3

Summary of branch circuit requirements for conductor and taps sizes, OCPD, and maximum load references.

CIRCUIT RATING	15 AMP	20 AMP	30 AMP	40 AMP	50 AMP
Minimum Size of Conductors:					
Copper Circuit Wires	14	12	10	8	6
Taps	14	14	14	12	12
OVERCURRENT PROTECTION	15 AMP	20 AMP	30 AMP	40 AMP	50 AMP
Outlet Device Type:					
Lampholder Type Permitted	Any Type	Any Type	Heavy Duty	Heavy Duty	Heavy Duty
Receptacle Rating*	Max. 15 amp	15 or 20 amp	30 amp	40 or 5 amp	50 amp
*Except for cord-connected electric-discharge light fixtures – See NEC 410.3(C)					
MAXIMUM LOAD	15 AMP	20 AMP	30 AMP	40 AMP	50 AMP
Permissible Load Reference:	NEC 210.23(A)	NEC 210.23(A)	NEC 210.23(B)	NEC 210.23(C)	NEC 210.23(C)

To keep things simple when you are sizing conductors for branch circuits, start by keeping in mind that the OCPD defines the circuit. If a 20-amp circuit contains eight AWG conductors because of voltage drop, it is still a 20A circuit. The rating of the branch circuit is determined by the size of the overcurrent protection device. An individual receptacle on a single branch circuit cannot have an amperage that is less than the rating of the OCPD, as specified in **NEC** Article 210.21(B)(1). Also remember that a single receptacle has only one contact device on its yoke, as described in Article 100, which means that you would treat a duplex receptacle as two receptacles.

Receptacle Quantity and Spacing

NEC 210.52 addresses receptacle spacing requirements for dwelling units. This is a subject that always seems to cause some confusion. The code states that "Receptacles shall be installed so that no point measured horizontally along the floor line in any wall space is more than 1.8 meters (6 feet) from a receptacle outlet." It goes on to describe what constitutes a wall space and that doorways do not count as part of the wall space.

FIGURE 2.4

Spacing outlets requires that receptacles be no more than 6 feet apart.

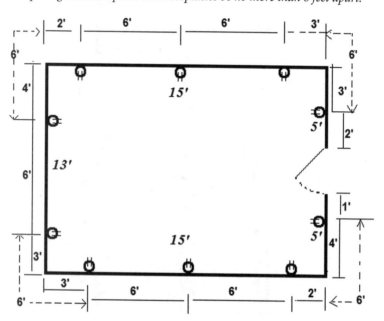

Trade Tip

The bottom line for receptacle spacing is that receptacles cannot be more than 12 feet apart along a wall line.

Kitchen receptacles have a couple of requirements. First, in Article 210.52(B)(3) you will find that wall countertop receptacles have to be supplied by at least two small-appliance branch circuits. Either of those circuits, or both of them, can supply receptacle outlets in the same kitchen and in other rooms, but they cannot supply more than one kitchen. Next, **NEC** 210.52(C) outlines that if a countertop is 12 inches wide or wider, then the receptacle outlets have to be installed no more than 24 inches apart. You should note the exception to this standard, which states that receptacle outlets are not required on the wall directly behind a range or sink. There are two diagrams provided in **NEC** Figure 210.52 that show how to determine where the 24-inch rule needs to be measured from where there is a sink or range in the counter.

Island countertops that measure at least 24 inches long and at least 12 inches wide also require at least one receptacle. However, if an island countertop has a range or sink installed and there is less than 12 inches behind it, then the code views the countertop as two separate countertop spaces. A peninsular countertop layout differs from an island countertop as shown in the figure below, but the size requirements of 24 by 12 inches and at least one receptacle are the same. The peninsular countertop is measured from the connecting edge of the main countertop.

The rest of Article 210 deals with outlets on wall spaces. Bathrooms in a dwelling unit must have a minimum of one receptacle outlet within 3 feet of the outside edge of each sink basin and on the wall that is adjacent to the sink or sink countertop. There must be at least one receptacle in dwelling-unit laundry rooms and laundry areas. The exception is laundry facilities in apartment buildings that are accessible to all the occupants and apartment buildings where laundry facilities are not installed or permitted. Attached and detached garages and basements all require at least one receptacle, and if a hallway is 10 feet long or longer, then it must also have at least one receptacle. The hallway measurement is the length along the centerline

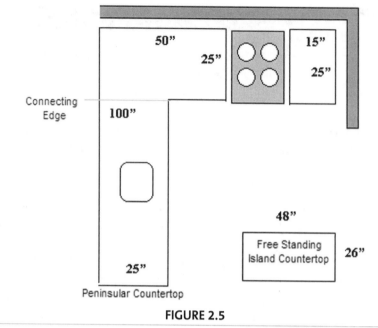

FIGURE 2.5

Countertops.

without the wall run being broken by a door. Finally, at least one wall-switched lighting outlet is required in every habitable room and bathroom.

FEEDERS

The definition of a feeder is that it is the circuit conductors that supply power to a branch-circuit overcurrent device or to a panel that contains the OCPD. The power can come from the service equipment, a separately derived system, or other power-supply source. So what is the main difference between a branch circuit and a feeder? A feeder runs between an overcurrent protection device (OCPD) at the supply and a downstream OCPD typically supplying a branch circuit. On the other hand, a branch circuit runs between the OCPD and an outlet, which is considered the final load. In other words, a feeder supplies power to a branch-circuit overcurrent protection device and that OCPD in turn powers the branch circuit. You need to remember that you don't size that branch circuit OCPD based on feeder calculations but rather on the branch-circuit load calculations and receptacle outlet requirements. **NEC** Article 215 is fairly short, because it only outlines the rules for installation, minimum size, and ampacity of feeders.

When you calculate feeder-circuit conductor loads, the minimum size, before any adjustments or correction factors are applied, needs to be no less than the noncontinuous load *plus* 125 percent of the continuous load. If the feeder conductors carry the total load supplied by service conductors that are 55 amperes or less, then the feeder conductor amperage cannot be less than the service conductor amperage. To size feeders that supply transformers, you need to add the transformer nameplate ratings together and make sure that the feeder conductor amperage is not less than that sum.

To protect feeder circuits, you must install overcurrent protection. If a feeder supplies a continuous load or a combination of continuous and noncontinuous loads, then the OCPD cannot be rated any less than the noncontinuous load plus 125 percent of the continuous load amperage. Both of the rating requirements have an exception for assemblies that are listed to operate at 100 percent of their ratings, including the overcurrent protection devices. In these cases, the ratings can be equal to the sum of the continuous and non-continuous loads.

A common neutral is permissible for two sets of 4-wire or 5-wire feeders, as well as for two or three sets of 3-wire feeders. Grounding is required for all feeders that supply branch circuits that require equipment grounding conductors. You can tap two-wire DC circuits or AC circuits with two or more ungrounded conductors. In this situation, you would tap from the ungrounded conductors of the circuits that have a grounded neutral conductor and use switching devices in each tapped circuit that has a pole in each ungrounded conductor.

The grounded conductor of a feeder circuit has to follow the same standards of identification as any grounded conductor, which means that it must have a continuous white or gray outer insulated covering or three continuous white stripes on an insulated covering that is not green. These requirements are listed in **NEC** 200.6. Equipment grounding conductors need to be identified in accordance with **NEC** 250.119.

Now that you have a good understanding of branch circuits and feeders, you should be able to understand how to calculate the load requirements that are covered in the next chapter.

3

Branch-Circuit, Feeder and Service Calculations

One of the most important aspects of planning any electrical installation is running accurate load calculations. There are a couple of things for you to remember when you are running these computations. First of all, if any of your calculations result in a fraction of an ampere that is .05 or less, you can simply drop that fraction. The other thing for you to remember is that load calculations are affected by demand factors, which is the ratio of the maximum demand of a system to the total connected load of that system. While different loads have different demand factors, the demand factor is always equal to or less than one. Article 220 provides the requirements for calculating branch-circuit, feeder, and service loads and is divided into five parts: general requirements for load calculations are in Part I (220.1 through 220.5); Part II (220.10 through 220.18) gives calculation provisions for branch circuits; the "standard method" for feeder and service-load calculation requirements is found in Part III (220.40 through 220.61); optional feeder and service-load calculation provisions are in Part IV (220.80 through 220.88); Part V (220.100 through 220.103) lists the calculation specifications for farm loads.

TERMS TO KNOW

Feeder: A circuit conductor that runs between a power-supply source and the final branch-circuit overcurrent device.

General Lighting Load: The portion of the total branch-circuit power draw that is used for general illumination.

Luminaires: Light fixtures.

Nameplate: A tag attached to an appliance, fixture, or motor that lists essential information regarding the item's electrical characteristics, such as the horsepower, amperage, wattage, and other ratings.

Volt-ampere: A measure of the apparent power in an alternating current circuit. Apparent power is the sum of real and reactive power.

STARTING WITH THE BASICS

Article 220 provides the calculation requirements for branch circuits and service feeder loads. This article begins with a load-calculation reference table, which is very handy for those times when you need to know more about sizing installations that are not necessarily commonplace for the average electrician. Here you will find load-calculation information for marinas and boatyards, listed in Article 555.12, calculations for phase converters, located in 455.6, and air-conditioning and refrigeration equipment requirements, provided in **NEC** Article 444, Part IV. The **NEC** also provides a convenience reference located in Annex D, which contains examples of load calculations.

Trade Tip

Think backwards and forwards. As you run calculations as they are laid out in Article 220, keep in mind that they are used in conjunction with provisions from other articles. Results from calculations in Parts III, IV, and V of Article 220, for example, are used with the provisions in 215.2(A)(1) to find the minimum feeder-circuit conductor size. The calculations in **NEC** 220 are also necessary for determining the minimum fuse or breaker size allowed for feeders as required by **NEC** 215.3.

Part I of Article 220 contains general calculation procedures, and Part II provides calculation provisions for branch circuits. Feeder and service calculation requirements are found in Parts III and IV; Part V contains the load requirements for farms. When you open up the code book to this article, you immediately know where to find calculation information, but knowing how to run the calculations and achieve accurate results is another matter altogether. Let's walk through the process, beginning with common receptacles.

GENERAL USE RECEPTACLES

Receptacle outlet load calculations for new installations are covered in **NEC** 220.14(I), (J), and (K). Section 220.14(I) lists the requirements for all types of receptacle outlets, with the except of outlets in dwellings, office buildings, and similar units. The standards for one-family, two-family, and multifamily dwellings are provided in **NEC** 220.14(J)(1) through (J)(3), and general lighting-load calculations are found in **NEC** 220.12. Unlike

FIGURE 3.1

A typical general-use receptacle.

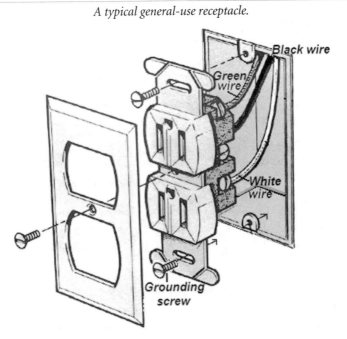

some other applications, there are no additional load calculations required for the receptacle outlets covered in 220.14(J).

All general-use receptacle outlets that are rated at either 15 or 20 amperes and are installed in dwellings are included as a part of the general lighting-load calculations. Let's look at an example. Assuming that a living room in a single-family home is 13 by 15 feet, first you have to space the receptacle installations in accordance with **NEC** 210.52. Because of the room's dimensions, nine receptacle outlets are required. Unlike receptacles for non-dwelling installations, which must be calculated at 180 volt-amperes each, dwelling receptacles do not require any additional load. The general lighting load for this room would need to be calculated in accordance with 220.12 and Table 220.12, which would be 3 volt-amps per square foot. This amounts to 13 by15 feet, which equals 195 square feet times 3 volt-amps, for a total of 585 units. The next step in the calculation process is not spelled out in the code article. Based on a normal residential service size of 120/240 volts, you need to complete the calculation by dividing the unit total by 120 (volts). This results in 4.875 (585 / 120 = 4.875), which should be rounded up to 5.

The various types of outlets specified in 220.14(J)(1) through (J)(3) include the following:

- J1: All general-use outlets rated at 20 amperes or less, which include receptacles that are connected to 20-ampere bathroom branch circuits as specified in **NEC** 210.11(C)(3)

- J2: Receptacles installed outdoors, as specified in **NEC** 210.52(E), as well as basement and garage receptacle outlets that are specified in **NEC** 210.52(G)

- J3: Lighting outlets in dwelling units, as described in **NEC** 210.70(A), as well as the outlets for guest rooms or guest suites in hotels and motels

Did You Know?

The load-calculation formula does not change when the number of receptacle outlets, specified in 220.14(J)(1) through (J)(3), is more than the required minimum.

> **Trade Tip**
>
> General lighting loads are based on the square footage (or square meters) of the occupancy space, and the calculation factors vary based only on the type of occupancy.

The interesting aspect about this section of the code is that the load calculation does not vary, even if you add more outlets for a space than are required by the standard. Going back to our 13-by-15-foot living-room example, if you added twice as many receptacles, the load for space would be calculated exactly the same as in our example: 195 square feet × 3 = 585/120 = 4.875.

CALCULATING BRANCH CIRCUITS

To determine the minimum number of 15- and 20-ampere branch circuits needed in dwellings for lighting and general-use outlets, you need to start with the floor area again. Begin by determining the square footage of the space and then, using Table 220.12, find the general lighting load required. The resulting lighting load is in amperes, which need to be divided by either 15 or 20 for the branch circuits. Let's look at a new example. Assume you have a single-family house with outside dimensions of 50 feet by 30 feet.

- STEP 1: Determine the square footage of the dwelling by multiplying the length time the width: 50 × 30 = 1500 square feet

- STEP 2: Calculate the general light load using Table 220.12, which indicates that the unit load for a dwelling unit is 3 volt-amperes per square foot. Multiply 1500 × 3 = 4500 volt-amperes.

- STEP 3: Confirm that the dwelling voltage is 120/240 volt and divide the minimum light load that you calculated by 120 volts: 4500 / 120 = 37.5, which should be rounded up to 38.

- STEP 4: To determine the minimum number of 15-ampere, 2-wire circuits that are required by the code, you need to divide the load you determined in above by 15 amps. This comes out to 38 divided by 15, which equals 2.533, which needs to be rounded up to 3. Now you have determined that at least three 15-ampere, 2-wire circuits are required for your 50-by-30-foot dwelling installation.

The good news is that you only need to remember this one procedure to also calculate the minimum number of 20-ampere general-purpose branch circuits you would need. The only part of the process that varies is that you use 20 as your factor instead of 15. So the calculation process ends up looking like this:

- $50 \times 30 = 1500$ square feet
- $1500 \times 3 = 4500$ volt-amperes
- $4500 / 120 = 37.5$ (round up to 38)
- 38 divided by **20** = 1.9, which needs to be rounded up to 2

So, based on our sample, the minimum number of 20-ampere, 2-wire branch circuits required for our dwelling is two. But you can't just stop there. At this point, you need to pause and ask yourself if there are any other code requirements that affect branch circuits. For example, we learned in the last chapter that you must have at least one 20-ampere laundry circuit and a minimum of two 20-ampere small-appliance branch circuits. Load calculations for small-appliance and laundry branch circuits are covered in **NEC** 220.52 and will be reviewed later in this chapter. This is a total of three branch circuits just to meet related code standards, and your minimum load calculation resulted in only two. How many do you use? The answer is three.

FIGURE 3.2

Small-appliance and laundry circuits must be included in load calculations.

If you stop to think about any other associated code requirements that you may have to consider, you will remember that bathroom branch circuits that supply receptacle outlets also have to be rated at 20 amperes. The good news is that no additional load calculation is required for the bathroom circuits, so you still only need three.

Several of the load conditions outlined in Article 220.14 include special calculation factors. Beginning with **NEC** 220.14 (A), you will find that special appliance loads have to be calculated based on the amperage rating of either the appliance or the load served. In 220.14(B) you will be referred to 220.54 for the load-calculation methods to use for electric dryers and to 220.55 for electric ranges and similar cooking appliances. You find in 220.14(C) that you need to go to **NEC** 430.22, 430.24, and 440.6 for the standards related to outlets for motor loads. Along similar lines to special appliance loads, **NEC** 220.14(D) standards indicate that outlets that supply lighting fixtures are to be calculated based on the maximum volt-ampere rating of the equipment and lamps that the fixture is rated for, while 220.14(E) goes on to require that heavy-duty lampholders need to be calculated at a minimum of 600 volt-amperes. Per **NEC** 220.14 (F), to determine the loads for signs and outline lighting, you need to start with a minimum of 1200 volt-amperes for each of the required branch circuits that are outlined in **NEC** 600.5(A).

Let's take a look at **NEC** 220.14(G) as an example of how you need to take a detailed look at the code in order to perform thorough load calculations, because some of the initial calculations in Article 220 are not stand-alone answers. For show windows in areas such as storefronts, the code indicates that you may choose between two methods. Either use the unit load per outlet method that is outlined in **NEC** 220.14 or allow 200 volt-amperes for every 1 foot of show window. Receptacle outlets that are not listed in 220.14(J) for dwelling units must be calculated at a minimum of 180 volt-amperes for every single or multiple receptacle on one yoke. If you had just one piece of equipment with a multiple receptacle that was made up of four receptacles, then you would run your calculation at no less than 90 volt-amperes per receptacle.

Now we will follow this calculation process all the way through. We'll assume that we have a show window that is 8 feet long. Step one begins with the fact that we know that in accordance with 220.14(G)(2) we must assume at least 200 volt-amperes for each foot of show window. Using just this information, the calculated load for this show widow is 8 feet × 200 volt-amperes, which equals 1600 volt-amperes. There are no specific requirements in this section to multiply continuous loads by 125 percent; however, there are requirements in other sections of other articles that have to be applied, and they are not referenced in this article.

Our next step is to determine if the show-window lighting in this store is a continuous load. For this, we need to refer to the definition in Article 100 of what constitutes a continuous load. Here we see that a continuous load is a load where the maximum current is expected to continue for three hours or more. Since almost every store is open for more than three hours a day, we know that the show-window lighting will be energized for at least three hours, which makes it a continuous load. We can quickly confirm this by referring to Example D3, in Annex D, which also confirms that show-window lighting is considered a continuous load.

In accordance with 210.19(A)(1) for branch-circuit conductors, a continuous load has to be multiplied by 125 percent. So we take the load of 1600 volt-amperes that we quickly calculated in our first step and multiply it by the required 125 percent: 1600 × 125 = 2000. Now we know that the branch-circuit conductors must be rated to carry at least 2,000 volt-amperes. To determine the minimum conductor ampacity, we next need to divide that by the 120-volt service size. So we take 2000 ÷ 120 = 16.7. But as an electrician, you know that we won't be using a circuit breaker this size and that we always need to round up to the next largest standard size. In this case we need a standard-sized 20-ampere circuit breaker. When you keep asking yourself if there are more standards to apply to any calculation, you can be assured that you are completely code-compliant. If you had not considered other sections in other articles, the branch-circuit conductors and overcurrent protection would have been the wrong size and in violation of the code (see 215.2(A)(1) for feeder conductors and **NEC** 215.3 for feeder overcurrent protection).

Banks and office buildings are not part of this calculation method, because they have their own standard listed in **NEC** 220.14(K). For these applications, you have to choose the larger of either the basic calculation method we have been practicing or 1 volt-ampere per square foot of floor area.

Trade Tip

Other sections of the **NEC** code that relate to branch circuits include:

- Air-conditioning and refrigeration equipment [440.6, 440.31, and 440.32]
- Appliances [422.10]
- Electric space-heating equipment [424.3]
- Information-technology/data-processing equipment [645.5]
- Motors [430.22]
- Signs [600.5]

FEEDER AND SERVICE LOADS

Understanding how to perform feeder and service load calculations is an important part of an electrician's professional task. Before we could calculate feeder and service loads, we needed to determine our branch-circuit loads. **NEC** 220.40 requires that the calculated load of a feeder or service not be less than the sum of the loads on the branch circuits supplied, as determined by Part II of Article 220, after any applicable demand factors permitted by Parts III or IV or required by Part V have been applied. Feeder and service load calculations include demand factors that should be applied to some of the branch-circuit load calculations that you have already run. Once you have included any of the applicable demand factors outlined in Part III or Part IV or required by Part V, you will be able to accurately total the branch-circuit loads and find the minimum required feeder or service size.

Fast Fact

Annex D is made up of numerous examples intended to explain how to calculate loads based on actual examples of various conditions such as multifamily dwellings and industrial feeders in a common raceway.

General Lighting Loads

You can actually lower the calculated lighting load of a feeder or service by utilizing the lighting load demand factors for certain occupancy types. Table 220.42 lists the demand factors that apply to the portion of the total branch-circuit load calculated for general lighting (illumination). Four types of occupancy lighting load conditions are provided: single-family and multifamily dwelling units, hospitals, hotels and motels, and storage warehouses. The hotel and motel standard includes apartment or rooming houses that do not have provisions for cooking by the tenants.

We will look at a single-family house with a calculated floor area of 4,000 square feet as an example of how to apply these demand factors. Table 220.42 indicates that the portion of the lighting load that is calculated at 100 percent is the first 3000 volt-amperes or less. The demand factor is 35 percent for the portion of the lighting load that ranges between 3,001 and 120,000 volt-amperes. That much is fairly clear, but now it gets a little more difficult. If you already know the comprehensive general lighting load, then you would just apply the rules in Table 220.42. But what if you only have the square footage of the house to go by? Let's go through the process of calculating the comprehensive general lighting load by using all of the code requirements and this table to apply demand factors for our 4000-square-foot house.

- STEP 1: Since you know the total floor area, begin by multiplying the square footage of 4000 by the 3 volt-amperes per square foot required in **NEC** Table 220.12: $4000 \times 3 = 12,000$. This is the base general lighting load.

- STEP 2: Next, we need to add the minimum of a single laundry circuit and two small appliance circuits that we know are required for a single-family dwelling - this was discussed in the section on **NEC** 210.11 in the previous chapter. Now that we have looked backwards, we have to read ahead to the circuit load requirements in **NEC** 220.52. Here we see that we need 1500 volt-amperes for each 2-wire, small-appliance branch circuit. So we multiply $1500 \times 2 = 3000$.

- STEP 3: **NEC** 220.52(B) requires that you also multiply each laundry branch circuit by 1500 volt-amperes, so we need to include another 1500 volt-amperes to our calculation.

- STEP 4: At this point, let's add together what we have so far: 12,000 (from Step 1) + 3000 (from Step 2) + 1500 (from Step 3) = 16,500.

- STEP 5: Use this total of 16,500 as your base to run the demand factors. The first 3000 is a straight 100 percent, which is simply 3000 volt-amperes. The remaining 13,500 (16,500 − 3,000 = 13,500) is subject to 35 percent demand: 13,500 × 0.35 = 4725.

- STEP 6: This means that the general lighting load for this one-family dwelling ends up being 3,000 (from Step 1) + 4725 (Step 5) = 7725 volt-amperes.

Now you have determined the most basic general lighting load for the house. However, there are additional electrical load elements that can affect service, circuit, and feeder sizing for any installation.

Service and Feeder Loads for Fixed Electrical Space Heaters

Article 220.51 simply states that any fixed electric space-heater loads are calculated at 100 percent of the total connected load. It adds the stipulation that under no circumstances can the feeder or service load current ratings be less than the largest supplied branch circuit rating. Whereas motor and motor-compressor loads must be calculated at 125 percent, fixed electric space-heating loads are simply based on 100 percent, but 100 percent of what exactly?

Assume you are reviewing plans for a small office and the specifications call for seven wall heaters. Each heater has a rating of 3000 watts at 240 volts. How much load will these heaters add to a 240-volt, single-phase service? If you just take **NEC** 220.51 on face value, then you would calculate the heaters at 100 percent (7 × 3000 = 21,000 watts) and call your formula done. However, we also need to determine the total current draw of the heaters. To do this, divide the total watts you just calculated by the 240 volts: 21,000 ÷ 240 = 87.5 Remember to round-up to 88 amperes. The calculated load for these seven 3000-watt, 240-volt heaters on a 240-volt, single-phase service is 88 amperes.

This process is fine if you just have standard heating units. But what if you are installing a heating and cooling combination unit? Heating units that are equipped with blower motors must be calculated in accordance with

220.51 and 220.50. To do this, let's assume we are going to install a heating/cooling package unit for a single-family dwelling. The specs for the unit indicate that the electric heater is rated 9.6 kW and 240 volts and the blower motor inside the package unit is a 1/2 horsepower, 240-volt motor. You job is to find out how much of a load this combination unit add to the house's 240-volt, single-phase service. Follow the steps to the process below:

- STEP 1: Convert 9.6 kW to basic watts using your basic electrical math that you had to learn before you could get your electrician's license. 9.6 × 1000 (kilo) = 9600 watts. Since the load needs to be eventually determined in amperes, you will need to convert the watts to amperes. Using Ohm's Law, this is done by dividing the watts by the amperes: 9,600 ÷ 240 = 40 amperes.

- STEP 2: Next you need to determine the full-load current in amperes of the 1/2-hp, single-phase, 240-volt motor. The information to do this is located in the section of the code in Article 430 that focuses on motors, motor circuits, and controllers. NEC Table 430.248 is a simple chart that lists the horsepower of A/C motors running at full-load currents and usual speeds with normal torque characteristics. In other words, the table is geared towards generic, average conditions. The motor we are using in our example is rated at 240 volts, but when you look at Table 430.248, you may get a little confused, because there is no column for motors rated at 240 volts. Just read the description of the table and you will see that the currents listed can be used for 110 volts to 120 volts and 220 volt to 240 volts.

- STEP 3: Based on Table 430.248, the full-load current for a 1/2-horsepower is 4.9 amperes. Unlike some of the calculations we have done previously, you need to use the exact voltage listed in the table; do *not* round up.

- STEP 4: Since we started with watts, we need to convert the 4.9 amperage rating to watts.

- STEP 5: Now add the amperes from the first step to those above: 40 + 4.9 = 44.9 amperes. This time you do need to round up to 45. This means that the total calculated ampere load for our heating and cooling package unit, including the blower motor, on a 240-volt, single-phase service is 45 amperes.

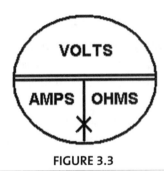

FIGURE 3.3

Ohm's law.

Keep in mind that unless this unit is the largest motor in the feeder or service load calculation, you do not need to include any additional demand factors. But just for the sake of knowing how to take these calculations to the next level, let's say that our 1/2-horsepower blower motor is the largest motor we have to include in the calculations for the house service. That being the case, the ampacity cannot be less than 125 percent of the full-load current rating *plus* the calculated load of the electric heat. Remember that our goal in all of this is to determine how much load all of the various elements in the house wiring layout will add to the dwelling's service requirements. To figure out how much load this package unit will add to our 240-volt, single-phase service, we need to multiply the motor's full-load current by 125 percent before adding it to the electric heat: 4.9 (from Table 430.248, Step 4 above) × 125 percent = 6.125. Next we have to add in the 40 amperes we calculated above to arrive at our total amperage: 40 + 6.125 = 46 amperes. You can see the difference in the calculated load when the blower motor is the largest motor in the service load calculation. It has increased from 45 to 46 amperes. This one-ampere change can be the difference in a service load that is code-compliant and one that is not.

FIGURE 3.4

NEC Table 430.248 may be used for 240 volts.

HORSEPOWER	115 VOLTS	200 VOLTS	208 VOLTS	230 VOLTS
1/6	4.4 amps	2.5 amps	2.4 amps	2.2 amps
1/4	5.8 amps	3.3 amps	3.2 amps	2.9 amps
1/3	7.2 amps	4.1 amps	4.0 amps	3.6 amps
1/2	9.8 amps	5.6 amps	5.4 amps	**4.9 amps**

Full-load amperage requirements for electric motors

Feeder and Service Loads verses Branch-Circuit Loads

Although fixed electric-space heating loads are calculated in accordance with 220.51 to determine the appropriate feeder and service loads, branch-circuit loads must be calculated in accordance with Article 424. Here we go, jumping around in the code again. By looking at Table 220.3, we know that we have to go specifically to Section 424.3 to find "Fixed Electric Space Heater Equipment, Branch-Circuit Sizing." We learned in **NEC** 210.19(A)(1) that branch-circuit conductors must have an ampacity that is not less than 125 percent of the continuous load. If we use our heating and cooling package unit as an example again, how would we now establish what minimum size 75-degree C branch-circuit conductors are required for the unit?

- STEP 1: The service load for this unit was calculated at 100 percent, but in accordance with **NEC** 430.22(A), the blower motor also had to be calculated at 125 percent. We went through this process and came up with a combined blower motor and heating element current of 44.9 amperes. Now we have to multiply that amperage by 125 percent: $44.9 \times 125 = 56.125$. (Back in the opening paragraph of this chapter, we pointed out that any fraction of an ampere less than .05 can be dropped, so this time we round *down* to 56.)

- STEP 2: We have now established that the 75-degree C branch-circuit conductors needed for our heating and cooling package unit must be at least 56 amperes.

- STEP 3: If you wanted to find out the minimum size 75-degree C conductors that would be required to feed this heating/cooling package unit, you would need to look in Annex B, Table B310.3, in the back of the code book. Here you will see that for 50 amperes at 75 degrees C you can use 8 AWG, but since we need 56 amperes, we have to jump up to the next level, which is 6 AWG copper conductors designed for use between 51 and 68 amperes.

No additional calculations are necessary as long as the heating and cooling unit provides the minimum circuit ampacity and maximum branch-circuit overcurrent protection that is shown on the nameplate. If our heating and cooling package unit was listed for a minimum/maximum fuse or

HACR-type breaker of 100 amperes on the nameplate, then the branch-circuit load would have to be protected by a 100-ampere fuse or HACR-type circuit breaker.

Small-Appliance and Laundry Loads

Small-appliance branch-circuit loads have to be included in the load calculations in order to determine the approved feeder and service loads for dwelling units. We included them already when we were calculating our comprehensive general lighting loads. **NEC 220.52.** requires 1500 volt-amperes for each two-wire small-appliance branch circuit. While there is a minimum number of these branch circuits required in **NEC 210.11**, there is no maximum number. In the kitchen, pantry, breakfast room, dining room, and other similar areas of a single-family house, the two or more 20-ampere small-appliance branch circuits required by 210.11(C)(1) have to serve all wall and floor receptacle outlets that are covered in **NEC** 210.52(A), as well as all the countertop outlets covered by 210.52(C) and all the receptacle outlets for refrigeration equipment as specified in **NEC** 210.52(B)(1).

If the load is subdivided through two or more feeders, the calculated load for each cannot be less than the 1500 volt-amperes for each two-wire small-appliance branch circuit. You are not required to include small-appliance branch-circuit loads on every feeder in a dwelling, so unless the feeder is going to supply small-appliance branch circuits, don't add small-appliance branch-circuit loads into your feeder calculations.

For example, let's say that a one-family dwelling will have three panelboards, not counting the service equipment. One of the panelboards is going to supply three small-appliance branch circuits, and the other two panelboards will not supply any small-appliance branch circuits. We have to multiply the three small-appliance branch circuits by the required 1500 amperes: 3 × 1500 = 4500. This means that 4500 amperes have to be included when we calculate the load for the panelboard with the three small-appliance branch circuits. But since the other two panelboards don't supply any small-appliance branch circuits, we will not include these loads in our calculations.

When you are looking at small-appliance circuits, remember that not all circuits in a kitchen are required to have 20-ampere circuits. Refrigeration equipment can be supplied from an individual branch circuit that is rated at 15 amperes or more according to **NEC** 210.52(B)(1), Exception # 2. But if the refrigeration equipment is not supplied by its own individual branch circuit, then it has to be supplied by a 20-ampere small-appliance branch circuit, and the load has to be calculated at the same 1500 volt-amperes as any of the other small-appliance circuits.

Calculations for laundry-circuit loads are a flat 1500 amperes, and only one circuit has to be provided. Again, you can add more circuits, but you always have to provide at least one.

Electric Clothes Dryers in Dwelling Units

The load for each household electric clothes dryer has to be the larger of either the nameplate rating or 5000 volt-amperes. Unlike the required minimum load of 1500 volt-amperes for a laundry branch circuit, there is no required minimum load if an electric clothes dryer is not going to be installed—for example, if a gas dryer is going to be installed.

When you calculate the load for a feeder or service when an electric clothes dryer will be installed, you have to include a load of at least 5000 volt-amperes. Let's look at an example of a duplex dwelling unit with a 4.5 kW (kilowatt) electric clothes dryer in one unit and a 5.6 kW dryer in the other unit and find the minimum service load for each dryer.

- STEP 1: First we have to convert the kilowatts to watts (or volt-amperes) by multiplying the kilowatts by 1000. In the first unit this would be: 4.5 $\times$ 1000 = 4500 volt-amperes. Right away you can see that at only 4500 volt-amperes, this does not meet the minimum load for a clothes dryer, so you would use the required 5000 volt-amperes.
- STEP 2: In the second unit, we multiply 5.6 $\times$ 1000, which equals 5,600 volt-amperes. Since this is larger than the minimum rating of 5,000 volt-amperes, you would use 5,600 as the minimum feeder load for this clothes dryer.

If there are five or more electric clothes dryers on a feeder or service, the load can be reduced. One to four dryers is calculated at 100 percent. In accordance with 220.54, you need to apply the demand factors in Table 220.54 to

Number of Dryers	Demand Factor %
1 - 4	100%
5	85%
6	75%
7	65%
8	60%

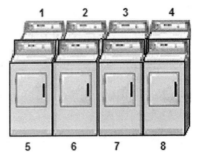

FIGURE 3.5

*Based on **NEC** Table 220.54, the demand factor for
8 electric clothes dryers is 60 percent.*

household electric clothes dryers. The process for an eight-unit multifamily unit where each unit has a 5.2 kW clothes dryer would work out as follows:

- STEP 1: Multiply the kilowatts of each dryer by 1000: 5.2 × 1,000 × 8 = 41,600.
- STEP 2: Next, find the demand factor percent to use for eight dryers in Table 220.54; it is 60 percent.
- STEP 3: Finally, you need to multiply the total dryer load by the demand factor: 41,600 × 60 percent = 24,960. The minimum clothes-dryer load for this multifamily dwelling is 24,960 volt-amperes.

Let's look at how to calculate the load for a multifamily unit with 20 electric clothes dryers that are each rated at 4.8 kW. First we multiply the 4.8 kW × 1000 and come up with 4,800 volt-amperes per dryer. Since this does not meet the minimum required by the code, we will need to start by using 5,000 volt-amperes for each dryer. Next, multiply by the number of

Trade Tip

Do not apply Table 220.54 demand factors to a laundromat in a multifamily dwelling, because it is possible that all dryers will be operating at one time.

> **CODE UPDATE**
>
> In the 2008 **NEC**, this formula is stated as: 47 minus 1% for each
> dryer exceeding 11. Both formulas yield the same demand factor
> percentages when calculating 12 to 23 dryers: the number of dryers
> exceeding 11 is 9 (20 − 11 = 9). The demand factor is 38 percent
> after subtracting 9 from 47 (47 − 9 = 38%).

dryers by the volt-amperes: 5,000 × 20 = 100,000. We have done this
process before, but now we need to find the demand factor percentage
from the formula listed for between 12 to 22 dryers. The 2005 edition of
NEC Table 220.54 used the following calculation formula: [% = 47 −
(number of dryers − 11)]. There were several parts of this equation which
were constants, which means that they did not change regardless of the
other parts of the formula. If you thought jumping ahead in the code has
been fun so far, just watch how we jump around to read this formula. Here
is how that calculation breaks down:

- Perform the part of the formula in the parentheses first: (number of dryers − 11). From the 20 dryers subtract the constant
  of 11. This equals 9.

- Now subtract that total of 9 from the constant of 47 to determine the demand factor: 47 − 9 = 38. This total is used in
  terms of the percentage, so when you read the formula if looks
  like this: %, which is 38, = 47 −9, which is the number of
  20 dryers − 11.

- Finally, you need to apply the demand factor to the total dryer
  load: 20 dryers × the minimum 5,000 amps = 100,000;
  100,000 × 38% = 38,000.

- The minimum clothes-dryer load for this multifamily dwelling
  is 38,000 volt-amperes.

FEEDER AND SERVICE-LOAD CALCULATIONS

Article 220.82 outlines dwelling-unit feeder and service loads with a connected load served by a single 120/240-volt or 208Y/120-volt 3-wire service or feeder conductors of 100 amperes or more. The total calculated load

is the result of adding the load formulas in 220.82(B) and 220.82(C). The first step is to take 100 percent of the first 10 kW and add to it 40 percent of the following loads:

- 3 volt-amperes per square foot of the dwelling unit for the general lighting and general-use receptacles. Remember that the dwelling-unit square footage should not include any garages, open porches, or unfinished spaces that will not be finished off in the future.

- 1500 volt-amperes for every 2-wire, 20-ampere small appliance branch circuit and laundry branch circuit.

- The nameplate rating of all of the appliances that will be fastened in place or permanently connected, such as ranges, clothes dryers, and water heaters.

- The nameplate amperage or kVA rating of any motors and of any low-power-factor loads.

Next you need to add the largest of the following kVA loads:

- 100 percent of any air-conditioning/cooling nameplate ratings.

- 100 percent of any heating nameplate rating if you are installing a heat pump without any supplemental electric heating.

- 100 percent of the nameplate rating of any heating systems, such as electric thermal storage units, that will have a typical continuous load that will be at the full value listed on the equipment nameplate.

- 100 percent of any heat-pump compressor nameplate rating and 65 percent of any supplemental central electric space heaters. There is an exemption to this part of the calculation, which is that if the heat-pump compressor is installed in a manner that keeps it from operating at the same time as the supplemental heaters, then it doesn't have to be added to the heaters as part of the total central heating load.

- 65 percent of the nameplate rating for any electric space heating if there are less than four individually controlled units.

- 40 percent of the nameplate rating for any electric space heating if there are four or more separately controlled units.

Feeder and Service Loads for Multifamily Units

As long as a service supplies three or more dwelling units and each unit has its own electric cooking and heating and/or air conditioning equipment and is only supplied by one feeder, you can use **NEC** Table 220.84 to calculate the feeder load. Figure 3.6 below illustrates these optional calculations. Demand factors listed in that table have to be applied to calculated loads that fall under the following criteria:

- 3 volt-amperes divided by the square feet for general lighting and use receptacles

- 1500 volt-amperes for every 2-wire, 20-ampere small appliance and laundry branch circuit

FIGURE 3.6

Optional load calculations are allowed for multifamily dwellings.

Number of Units	Demand Factor
3 to 5	45%
6 or 7	44%
8 to 10	43%
11	42%
12 or 13	41%
14 or 15	40%
16 to 17	39%
18 to 20	38%
21	37%
22 or 23	36%
24 or 25	35%
26 or 27	34%
28 or 30	33%
31	32%
32 or 33	31%
34 to 36	30%
37 or 38	29%
39 to 42	28%
43 to 45	27%
46 to 50	26%
51 to 55	25%
56 to 61	24%
62 and over	23%

- The nameplate rating for all permanently connected appliances on a specific circuit, such as ranges, clothes dryers, water heaters and space heaters
- The nameplate amperage or kilovolt amperage of all motors and low-power-factor loads
- The air-conditioning load or space heating load, whichever is the largest

If you are working on other types of installations such as schools, restaurants, or buildings on farms, then you need to refer to **NEC** 220.86, 220.87, and **NEC** V, respectively. Determining the required branch-circuit feeders and calculating service loads is the jumping-off point for installing electrical services and proper overcurrent protection, which are covered in the next chapter.

CHAPTER

4

Services and Overcurrent Protection

WHAT YOU NEED TO KNOW

Article 230 of the **NEC** describes the standards for providing electrical-supply services to various types of buildings and equipment. It ties in to NEC 240, which covers over-current protection and transformers, since there are many situations in which a transformer may be the first unit of equipment that is connected to the electrical service. Article 230 starts out with a convenient graphic that illustrates an electrical service and references the various parts of the code that provide regulations for each segment of the service. In the electrical field, grounding and bonding and overcurrent protection are probably the two most essential and important protective principles. Overcurrent protection is critical for overall personal safety in terms of a number of hazardous conditions that could be caused by materials igniting due to improper overload or short-circuiting. Additionally, OCPD guards against explosive ignition and flash hazards from inadequate voltage-rated or improper interrupting-rated overcurrent protective devices. Once you are clear on how to protect current flow, you will also need to know the basics of how transformers work so that you can install and connect to them properly. First we will cover how to provide the electrical service to a location and then how to protect dwellings and equipment from the prospective hazards of that electricity.

SERVICE

Service Drops and Clearances

Service conductors that are installed as open conductors or multiconductor cable without an outer jacket have to clear buildings by at least 3 feet from any windows or doors that can be opened or any porches, balconies, exterior stairs, or similar structures. Overhead service conductors can't be installed under or in the way of building openings such as loading dock doors, where materials being moved in or out could come in contact with the service. And, not that many professional electricians would considering doing this, but **NEC 230.10** lets you know that you are not allowed to use trees or other types of vegetation to support overhead service conductors.

Part II of Article 230 provides the basics regarding overhead service drops as follows:

- With the exception of the grounded conductor of a multiconductor cable, which can be bare, all other service conductors have to have an insulated covering.

- Conductors need to be adequately sized to serve the loads that you calculated in Chapter 3 for each dwelling unit.

- Service conductors have to be at least 8 AWG copper or 6 AWG aluminum/copper-clad aluminum. Limited load services of a single branch circuit for installations such as small polyphase power are the only exception to this requirement, and these can not be any smaller than 12 AWG hand-drawn copper.

- Conductors have to clear roofs by 8 feet, unless the voltage between conductors is less than 300 volts, in which case there are a number of exceptions provided in **NEC 230.24(A)** regarding clearances.

- The standard vertical clearances for service-drop conductors are:
  - Clearance from the ground of 10 feet
  - Over residential property or driveways 12 feet
  - 15 feet if the voltage in the 12-foot classification exceeds 300 volts to ground
  - 18 feet over public streets, roads, parking lots or alleys

Service Conductors and Equipment

A service drop or lateral can only supply one set of service-entrance conductors. **NEC** 230.42 refers you to Article 220 in order to establish the acceptable loads for these conductors and to **NEC** 310.15 to ascertain the required amperages. It goes on to list two calculation requirements for determining the ampacity of the service-entrance conductors before any adjustment or correction factors are applied. The ampacity can not be less than:

- The total of all noncontinous loads plus 125 percent of the continuous loads
- The total of all of the noncontinous loads plus the continuous load if the service-entrance conductors are terminated in an overcurrent-protection device that is rated to operate at 100 percent of the conductor's rating

The service equipment has to be either enclosed or guarded per **NEC** 230.62 (A)(1) and (A)(2). This means that any energized parts have to be enclosed so they cannot be exposed to accidental contact; if they are not enclosed, they must be installed on a panelboard, switchboard, or control board in a suitable location as described in **NEC** 110.27.

Fast Fact

"Suitable" means a room or vault that is only accessible by a qualified person or an elevated platform more than 8 feet off the floor.

You have to provide a means to disconnect the service equipment that is readily accessible and permanently marked to identify it as a service disconnect, per **NEC** 230.70. The service disconnect for each service or set of service-entrance conductors cannot be made up of more than six switches or sets of circuit breakers or a combination of the two, and no more than these six disconnects can be grouped together in any one location per service. A multipole disconnect that consists of two or three single-pole switches or breakers that operate individually is permitted on multiwire circuits as long as there is one pole for every ungrounded conductor and all have handle ties or a master handle that can be operated by no more than six hand movements to disconnect all the service conductors, as per **NEC** 230.71.

If the installation method for a service disconnect does not disconnect the grounded conductor from the premise wiring, then you will need to provide a terminal or bus with pressure connectors that attach all of the grounded conductors. You may also use a marked multisection switchboard. Obviously, the service disconnect means can-not be rated at less then the load it would carry, and it can never be rated lower than any of the following:

- 15 amperes minimum for disconnects to a service for the limited loads of single branch circuits
- 30 amperes minimum for disconnects to a service for no more than two 2-wire branch circuits
- 100 samperes minimum for disconnects to a single-family dwelling service
- 60 amperes minimum for all other service disconnect means

There are limited types of equipment that can be connected to the supply side of a service disconnect means, as described in **NEC** 230.82. These include cable limiters, instrument transformers for current and voltage, impedance shunts, arrestors, and ground-fault protection devices that are installed as part of the listed equipment but only if a suitable overcurrent protection and disconnect means is provided. In addition, ground-fault protection of equipment is required to be directly connected to ground for any solidly grounded wye electrical services between 150 volts to ground and 600 volts phase-to-phase for any service disconnect that is rated at 1000 amperes or more.

OVERCURRENT PROTECTIVE DEVICES

Overcurrent protective devices, or OCPDs, are typically used in main service disconnects and the feeders and branch circuits of electrical systems for residential, commercial, institutional, and industrial premises. OCPDs are meant to protect against the potentially dangerous effects of overcurrents, such as an overload current or a short-circuit current, which often is referred to as a fault current. Equipment damage,personal injury, or even death can result from the improper application of a device's voltage rating, current rating, or interrupting rating. Something as simple as a circuit breaker can protect against this damage, but if a fuse or circuit breaker doesn't have an adequate voltage rating, it can rupture or explode while attempting to stop fault currents beyond their interrupting ratings.

Fuses and circuit breakers are intended to provide overload and short-circuit protection and are located at the line side of the circuits, such as

FIGURE 4.1

Typical overcurrent-protection devices.

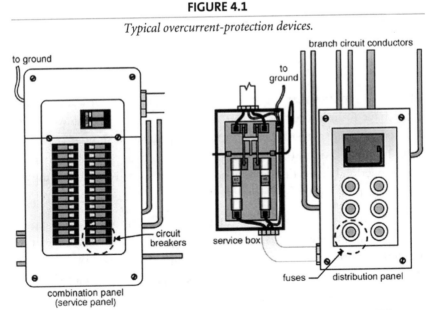

OVERCURRENT DEVICES CAN BE CIRCUIT BREAKERS OR FUSES

heating and lighting branch circuits, that are being protected. **NEC** 240.3 requires branch circuit, feeder, and service conductors to be protected against overcurrent based on their ampacities as specified in Table 310-16. However, this section also contains twelve rules that modify the general requirement and allow conductors not to be protected in accordance with their ampacities under the following conditions:

- Power-loss hazard
- Devices rated at 800 amperes or less
- Tap conductors
- Motor-operated appliance circuit conductors
- Motor and motor-control circuit conductors
- Phase converter supply conductors
- Air-conditioning and refrigeration equipment circuit conductors
- Transformer secondary conductors
- Capacitor circuit conductors
- Electric welder circuit conductors
- Remote-control, signaling, and power-limited circuit conductors
- Fire-alarm-system circuit conductors

NEC 240.4 requires that conductors, except for flexible cords and fixture wires, have to be protected against overcurrent in accordance with their ampacities. There are a number of conductor protection standards listed in **NEC** 240.4(A) though (G). For example, 240.4(A) does not require conductor overload protection for fire pump circuits because interrupting the electrical circuit would create a safety hazard.

NEC 240.4(B) allows you to use the next higher standard OCPD rating above the amperage of the conductors being protected for OCPDs that are 800 amperes or less, but only if the conductor ampacity doesn't already correspond to a standard overcurrent-protection device size. The next higher standard size cannot exceed 800 amperes. Additionally, the

2 AWG Conductors 75° C

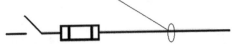

Based on NEC Table 310.16, 75° C 2 AWG copper conductors have an ampacity of 115 amps, but a fuse or circuit breaker could be sized as high as the next standard size of 125 amps

FIGURE 4.2

Sizing fuses or circuit breakers.

protected conductors cannot be part of a multi-outlet branch circuit that supplies portable cord-and-plug outlets. Additionally, the amperage of the conductors can't be the same as the standard ampere rating of a fuse or circuit breaker without overload trip adjustments above its rating.

NEC 240.4(C) requires that the conductor amperage has to be at least equal to if not greater than the rating of the overcurrent device if it is rated over 800 amperes.

NEC 240.4(D) establishes conductor size-limitation standards. The overcurrent protection device can not exceed the following:

- 15 amperes for 14 AWG
- 20 amperes for 12 AWG
- 30 amperes for 10 AWG
- 15 amperes for 12 AWG and 25 amperes for 10 AWG aluminum and copper-clad aluminum after any correction factors for ambient temperature and number of conductors have been applied

Did You Know?

The ampere rating of a fuse or circuit breaker is the maximum amount of current that it can safely carry without opening.

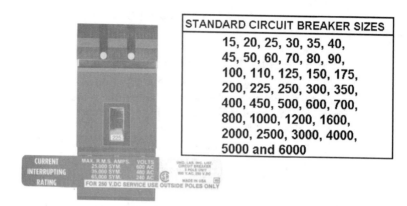

FIGURE 4.3

Standard circuit-breaker sizes.

NEC 240.6 lists the standard ampere ratings for fuses and inverse time circuit breakers. The standard ampere ratings for fuses are 1, 3, 6, 10, and 601. Standard circuit-breaker sizes are listed in the figure above.

Typically, the amperage rating of a fuse or a circuit breaker is based on 125 percent of the continuous load current. And, since conductors are also typically calculated at 125 percent of the continuous load current, the conductor ampacity wouldn't be exceeded. For example, a 40-ampere continuous load conductor multiplied by 125 must be rated to carry 50 amperes, so a 50-ampere circuit breaker is the largest that should be used.

There are certain circumstances when you can use a fuse or circuit breaker ampere rating that is greater than the current-carrying capacity of the circuit.

FIGURE 4.4

Multiplying the conductor size and the circuit-breaker size by 125 percent.

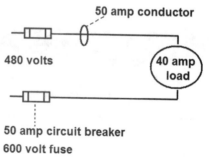

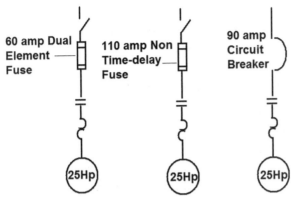

**460 volt 3-phase Motor rated @ 25Hp, 34 amps
50 amp Conductor rated for 8 AWG, 75° C**

FIGURE 4.5

*This motor circuit would allow fuses sized at exactly
1.75 × 34 amperes = 59.5 amperes.
Use the next standard size of 60 amperes.*

A typical example would be motor circuits, because dual-element, time-delay fuses are generally allowed to be sized up to 175 percent or the next standard of the motor full-load amperes. The figure below illustrates a 1.15 SF motor.

The conductors for this example would be sized based on **NEC** 430.22, which would be 34 × 1.25 = 42.5-ampere minimum. Based on **NEC** 310.15 and Table 310.16, you would use an 8 AWG, 75-degree C conductor, assuming the terminations are rated for 75-degree C conductors. Because the required overload relay or "heater" will be sized at 125 percent or less of the motor full-load amperes and provide the overload protection for the circuit, the 60-ampere time-delay fuse could be used. The conductor is

CODE UPDATE

Location of overcurrent devices. A new subsection in the 2008 **NEC** for 240.24 outlines locating overcurrent devices in stairways. Overcurrent devices are now prohibited over steps in stairways but can be located over landings adjacent to stairways.

also sized at 125 percent of the motor full-load amperes, and the overload relay is intended to protect the conductor from overloads since it is sized at or lower than the conductor capacity.

Voltage Ratings for Circuit Breakers and Fuses

The voltage rating of a fuse or circuit breaker is the highest voltage that is capable of safely interrupting under all overload and short-circuit conditions for which it is rated to interrupt. In order for an overcurrent protective device to operate correctly based on its voltage rating, that rating has to be equal to or greater than the system voltage. For example, you can use a 600-volt fuse or circuit breaker in a 575 V, 480 V, 208 V, or 120 V circuit, but a 250-volt fuse or circuit breaker would not meet **NEC** standards if you wanted to install it for a 480 V or 277 V circuit. There are two physical aspects to the operation of voltage-rated overcurrent-protection devices. The first is that it must have sufficient creepage and clearance distances to eliminate any conductive path or flashover between conductive parts of different phases, such as phase to neutral or phase to ground.

Next, the OCPD voltage rating is also a function of its capacity to open a circuit under an overcurrent condition, and it determines the ability of the overcurrent-protection device to suppress and snuff out the internal arcing that occurs during an overcurrent condition. If an OCPD with a voltage rating lower than the circuit voltage were to be installed, then its arc suppression and ability to extinguish the arc would be impaired and it might not clear the overcurrent safely.

FIGURE 4.6

Creepage and clearance distances at the terminations of a disconnect.

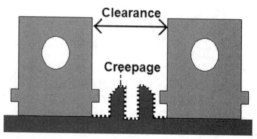

Did You Know?

Overcurrent-protection devices can be rated for AC voltage, DC voltage, or both, and often an AC/DC voltage-rated OCPD will have an AC voltage rating that is different from its DC voltage rating.

There are two types of AC-rated overcurrent-protection devices: straight voltage-rated and slash voltage-rated. All fuses are straight voltage-rated, but some circuit breakers are slash voltage-rated at 480/277 volts, 240/120 volts, or 600/347 volts. There are a number of types of circuit breakers and other multiple-pole, mechanical overcurrent-protective devices, such as self-protected starters and manual motor controllers, that may have a slash voltage rating instead of a straight voltage rating.

Most low-voltage power-distribution fuses have 250-volt or 600-volt ratings. **NEC** 240.61 allows fuses rated 600 V or less to be used for voltages below their rating. **NEC** 240.60 (A)(2) allows 300-volt-rated cartridge fuses to be used on single-phase line-to-neutral circuits that are supplied from 3-phase, 4-wire, solidly grounded neutral sources as long as the line-to-neutral voltage does not exceed 300 volts. This means that 300-volt cartridge fuses have to be used on single-phase 277 V lighting circuits.

Most circuit breakers that are used in low-voltage power-distribution installations have a voltage rating of either 125 volts, 250 volts, 480 volts, or 600 volts. **NEC** 240.83 (E) prohibits the voltage rating of circuit breakers from being less than the nominal system voltage. **NEC** 240.85 lists special requirements for the voltage rating of circuit breakers such as slash ratings. For example, a circuit breaker with a straight voltage rating, such as 240 volts or 480 volts, can be used on a circuit where the nominal voltage between any two conductors is not more than the circuit breaker's voltage rating.

Trade Tip

You cannot use a two-pole circuit breaker to protect a 3-phase, corner-grounded delta circuit unless the circuit breaker is marked 1Φ-3Φ.

Fast Fact

The most suitable application of series-rated combinations is for branch-circuit lighting panels.

Series Ratings

NEC 240.60(C) requires that that the minimum interrupting rating of branch-circuit cartridge fuses is 10,000 amperes, and **NEC** 240.83(C) establishes the minimum interrupting rating of branch-circuit circuit breakers as 5,000 amperes. These minimum interrupting ratings don't apply to supplemental protective devices such as glass tube fuses.

In **NEC** 240.86 there is a provision for fuses or circuit breakers to protect downstream circuit breakers if the available short-circuit current exceeds the downstream circuit breaker's interrupting rating. This is termed a series-rated combination, series rating, or series combination rating; however, in the field series-rated combinations should be used sparingly.

APPLYING THE CODE

Let's combine the concepts and requirements that we have covered so far to review the rules for conductor sizing and protection:

- STEP 1: Size the overcurrent-protection device in accordance with Sections 210.20(A), 215.3, and 384.16(D). These three **NEC** rules require that the breaker or fuse overcurrent-protection device be sized at no less than 100 percent of the noncontinuous load plus 125 percent of the continuous load. (Section 240.6(A) contains the list of standard-size overcurrent-protection devices.)

- STEP 2: Select the proper conductor size in accordance with Sections 110.4(C) and **NEC** 210.19(A), 215.2, and 230.42(A), which require that the conductor be sized no less than 100 percent of the noncontinuous load plus 125 percent of the continuous load. In addition, Section 110.14(C) requires you to consider the temperature rating of the equipment terminals when you size the conductors. In order to do this, you need to size the circuit conductors according to the 60-degree C column of Table 310-16 for equipment that is rated 100 amperes and less. Equipment rated over 100 amperes needs to be sized based on the 75-degree C column of Table 310-16.

■ STEP 3: Now that you have selected the proper conductors to use, you have to protect them against overcurrent in accordance with **NEC** 240.3, which requires the branch circuit, feeder, and service conductors to be protected in accordance with their ampacities as specified in Table 310.16. Remember when you do this that section 240.3(B) allows you to use the next size up if the conductors are not part of a multi-outlet branch circuit that supplies receptacles as long as the ampacity of the conductors doesn't correspond with the standard ampere rating of a overcurrent-protection fuse or a circuit breaker listed in **NEC** 240.6(A). Also, the next higher standard rating cannot exceed 800 amperes.

Assume that you have 19 kVA of nonlinear loads with 75-degree C terminals and that the branch circuit is supplied by a 208/120-volt, 4-wire, 3-phase setup. Let's walk through the steps to determine which size of branch-circuit overcurrent-protection device and conductor (THHN) are required for this installation based on the process we just reviewed:

■ STEP 1: Size the overcurrent protection device in accordance with 210.20(A) and 384.16(D). The first thing that you need to do is to convert the nonlinear load from kVA to amperes:

$$\text{amperes} = \text{VA}/(\text{Volts} \times 1.732)$$

$$\text{amperes} = 19,000/(208 \text{ volts} \times 1.732),$$

$$\text{amperes} = 52.74 \text{ amperes}$$

Round up to 53 amperes

The branch-circuit overcurrent-protection device has to be sized based on at least 125 percent of the 53 amperes:

$$53 \text{ amperes} \times 125\% = 66 \text{ amperes}$$

You need a minimum 70-ampere overcurrent-protection device, based on **NEC** 240.6(A).

■ STEP 2: Select the proper conductor size in accordance with Sections 110.4(C) and **NEC** 210.19(A), 215.2, and 230.42(A). **NEC** 210.19(A) also requires that the branch-circuit conductor be sized no less than 125 percent of the continuous load:

$$53 \text{ amperes} \times 125\% = 66 \text{ amperes}$$

You need to pick the conductor based on the 75-degree C terminal temperature rating of the equipment terminals. No. 6 THHN has a rating of 65 amperes at 75 degreesC, so you cannot use that size. You need to go to the next size up:

Use #4 THHN, which has a rating of
85 amperes at 75°C.

■ STEP 3: Protect the #4 THHN conductor against overcurrent in accordance with **NEC** 240.3. First, you need to confirm that the #4 THHN is properly protected against overcurrent by the 70-ampere overcurrent-protection device listed. To do this, you need to consider the fact that you have a 4-wire, 3-phase service. Since there are more than three current-carrying conductors in the same raceway, you have to adjust the #4 THHN conductor ampacity to the amperage listed in the 90-degree C column of **NEC** Table 310.16.

Corrected ampacity for #4 THHN = ampacity × Note 8(A). The adjustment-factor-corrected ampacity of #4 THHN = 95 amperes × 80%:

Corrected ampacity #4 THHN = 76 amperes

For #4THHN rated at 76 amperes after ampacity factors, the proper overcurrent-protection device will need to be rated at 70 amperes in order to comply with the general requirements of **NEC** 240.3.

Let's work through another example using a 184-ampere feeder continuous load on a panelboard with 75-degree C terminals that supplies nonlinear loads. We will assume that the feeder is supplied by a 4-wire, 3-phase, wye-connected system and need to determine which size feeder overcurrent-protection device and THHN conductor are required to meet code standards.:

- STEP 1: Size the overcurrent-protection device in accordance with **NEC** 215.3 and 384-16(D). The feeder overcurrent-protection device be sized at least 125 percent of 184 amperes:

    184 amperes $\times$ 125% = 230 amperes

    According to **NEC** 240.6(A) you must use a minimum 250-ampere overcurrent-protection device.

- STEP 2: Select the conductor that will comply with **NEC** 110.14(C) and with **NEC** 215.2, which require the feeder conductor to be sized no less than 125 percent of the continuous load:

    184 amperes $\times$ 125% = 230 amperes

    You then need to select a conductor according to the 75-degree C temperature rating of the panelboard terminals:

    #4/0 THHN has a rating of 230 amperes at 75°C

- STEP 3: Protect the #4/0 conductor against overcurrent based on the requirements in **NEC** 240.3. Start by verifying that the #4/0 THHN conductor would be adequately protected by a 250-ampere overcurrent-protection device. Since our sample installation is a 4-wire, 3-phase, wye-connected system, you know you will have more than three current-carrying conductors in the same raceway, so you have to correct the #4/0 THHN conductor ampacity by using the 90-degree C column of **NEC** Table 310-16:

    corrected ampacity for #4/0 THHN =
    ampacity $\times$ Note 8(A)

    The adjustment-factor-corrected ampacity for #4/0 THHN = 260 amperes $\times$ 80%:

    corrected ampacity for #4/0 THHN = 208 amperes

The #4/0 THHN that is rated at 208 amperes after you applied the ampacity correction would not be properly protected by a 250-ampere overcurrent protection-device, because the "next size up rule" in **NEC** 240.3(B) would only permit a 225-ampere OCPD on your 208-ampere conductor. This means that you have to increase the conductor size to 250 kcmil in order to comply with the overcurrent protection rules of **NEC** 240.3

By now, you can see that you may need to apply numerous standards to any given electrical installation in order to be completely code-compliant.

Grounding and Bonding

WHAT YOU NEED TO KNOW

There are two vital reasons for using grounding equipment. One is to protect people who come in contact with energized metal parts due to a ground fault, and the other is to ensure that the fault is quickly resolved before a fire develops. In the previous chapter, we reviewed how overcurrent-protection devices work and indicated that they must open quickly in order to clear a ground fault. In order for the OCPD to open, a grounding path must exist that has a low enough impedance to allow the ground-fault current to reach a level that is greater than the overcurrent-protection-device rating. When metal parts are permanently joined together, creating a conductive path for electricity, continuity and the capacity to safely conduct the electricity result. This is the purpose and general definition of bonding. Bonding metal parts to each other and then to the system provides the process required for overcurrent-protection devices to do their job. This is the topic of **NEC** 250.

TERMS TO KNOW

Many of the definitions that were listed in Article 100 come into play in the various standards and requirements concerning grounding and bonding, and Article 250 provides some additional definitions. Let's take a minute to give them a quick review.

Bonded (Bonding): Connected in a manner so as to establish electrical continuity and conductivity.

Bonding Jumper, Main: The connection between the grounded circuit conductor and the equipment grounding conductor at an electrical service.

Equipment Grounding Conductor (EGC): A conductive path that is installed to connect the normally non-current-carrying metal parts of equipment together and also to the system grounded conductor, the grounding electrode conductor, or both.

Ground: The earth.

Grounded (Grounding): Connected to the ground or to conductive material that extends the ground connection.

Grounded Conductor: A system or circuit conductor that is intentionally grounded.

Ground Fault: The <u>unintentional</u> electrical conductor connection between an ungrounded circuit conductor and normally non-current-carrying conductors, metallic enclosures, metallic raceways, metallic equipment, or the earth.

Ground-Fault Current Path. An electrically conductive path from the point of a ground fault on a wiring system through normally non-current-carrying conductors, equipment, or the earth to the electrical supply source.

Grounding Conductor: A conductor that is used to connect equipment or the grounded circuit of a wiring system to a grounding electrode.

Grounding Electrode Conductor: A conductor that is used to connect the system grounded conductor the equipment to the grounding electrode or to a point on the grounding electrode system.

UNDERSTANDING GROUNDING

Grounding is the process of intentionally connecting current-carrying conductors to the earth. In AC-premises wiring systems, **NEC 250.24** requires that this ground connection be made between the line side of the service equipment and the supply source, such as a utility transformer. There are three basic reasons for grounding:

- Limit voltage surges caused by lightning, utility-system operations, or accidental contact with high-voltage lines
- Provide a connection to the earth that can stabilize voltage under normal operating conditions
- Facilitate the operation of overcurrent devices such as circuit breakers, fuses, and relays under ground-fault conditions

NEC 250.4 (A) lists grounded systems that are required to be connected to the earth. These include the following:

- In order to limit the voltage to ground, any non-current-carrying materials that are conductive and enclose electrical equipment or conductors have to be connected to the earth.
- To establish an effective ground-fault current path, those same non-conductive-current-carrying materials have to be bonded together and to the electrical supply source.
- Additionally, any electrically conductive materials that are likely to become energized have to be bonded in the same manner.
- Any electrical equipment, wiring, and conductive materials that are likely to become energized have to be installed in a way that creates a permanent, low-impedance circuit that will allow OPCDs to work properly and to carry the maximum ground-fault current from any point on the wiring system to the electrical supply source. For this purpose, the earth is not considered an approved ground-fault current path.

Ground Faults versus Short Circuits

In order to comply with these standards, it helps to know what a ground fault is. It is similar to a short circuit in a way. A short circuit is an unintentional connection between two conductors; it can be either phase-to-phase or phase-to-neutral. A ground fault is an unintentional connection between an ungrounded, or "phase," conductor and a conductive material such as a metal enclosure or raceway or a metal equipment frame.

Short circuits are not all that common, because two insulation failures would have to occur for the unintentional connection to take place. In this kind of situation, the required overcurrent device that protects the circuit opens quickly. Ground faults, however, are more common, they only require a single insulation failure. A ground fault can be much more destructive than a short circuit. If you do not install a code-compliant ground-fault current return path, the result could be a high-impedance arcing fault that lasts a long time without causing a circuit breaker to trip or a fuse to blow. To compensate for this type of problem, arc-fault circuit interrupters (AFCIs) are required.

The **NEC** makes a distinction between a ground-fault current path and an effective ground-fault current path. A ground-fault current path is any conductive material that fault current can flow through. These materials are not limited to conductors or equipment and could include water lines

CODE UPDATE

A new Fine Print Note in the 2008 edition of **NEC** 250.4 advises you about the benefits of reducing the length of the grounding electrode conductor by calling your attention to Sections 800.100(A)(5), 810.21(E), and 820.100(A)(5), which require grounding conductors to be run as short and as straight as possible. This results in the most effective path to the earth for line surges caused by lightning events.

> **Trade Tip**
>
> Section 250.4(A)(5) requires that an effective ground-fault current path must be:
>
> - Permanent
> - Low-impedance
> - Able to carry the maximum ground-fault current likely to be imposed on it from any point on the wiring system

or gas pipes, air ducts, communications wiring, fences, rain gutters, and so on. An effective ground-fault current path is an intentionally designed and constructed low-impedance route meant to carry ground-fault current and facilitate overcurrent-protective-device operation. In other words, a ground-fault current path can be accidental, but an effective ground-fault current path is always an intentional and critical element of an electrical grounding and bonding system.

NEC 250.4(A)(5) prohibits you from using or considering the earth as an effective ground-fault current path. There is an important reason for this, which is that the earth's resistance is so high that hardly any fault current returns to the electrical-supply source through it. For this reason, the code requires that equipment grounding conductors have to be run with circuit conductors.

Objectionable Current

Simply put, there are good kinds of current and objectionable types of current. **NEC** 250.6 provides requirements for preventing the flow of objectionable current over grounding conductors. What it does not spell out clearly is the level of current that would deem it objectionable in any given situation. However, when you combine the definitions of grounding found in **NEC** 100 and **NEC** 250.2, you can get a clear idea of what makes current objectionable and how to prevent it. Grounding conductors are not meant to be used as circuit conductors for functions other than those specifically laid out in **NEC** 250.4. In most electrical installations, current will flow through capacitive coupling, and any possible or associated shock or fire hazard for those installations helps to clarify which levels of

Did You Know?

An example of an objectionable current would be a rise in potential on exposed metal parts that are not designed or meant to be energized, because this kind of situation could produce an electric shock or result in a fire hazard. Another scenario would be excessive current that would interfere with the proper operation of electrical or electronic equipment.

current are unsafe or "objectionable." Essentially, objectionable current is simply any level of current in an electrical installation that would pose an electric shock or fire hazard or hinder the ability of the grounding system to perform its intended functions.

NEC 250.6 requires that materials and equipment be installed and arranged in a way that prevents objectionable current from flowing over the grounding conductors or grounding paths. However, if the use of multiple grounding conductors would actually induce or allow objectionable current, then **NEC** 250.6(B) allows you to make several alterations as long as there is an effective ground-fault current path and a path for fault current is provided. These permitted adjustments include the following steps:

- You can disconnect one or more grounding connections, just not all of them
- You are allowed to change the locations of grounding conductors
- You could take other suitable corrective actions.
- You could even interrupt the continuity of the grounding conductor or the conductive pathway interconnecting the grounding conductors.

Grounding AC Systems

Alternating-current systems are required to be grounded per **NEC** 250.20. For systems that are less than 50 volts, you need to ground installations that are supplied by transformers that exceed 150 volts to ground. You also need to ground systems with a transformer supply that is ungrounded or installed as overhead conductors outside of a building. If the system has a supply between 50 and 1000 volts, you will have to ground it in any of the following ways:

- So that the maximum voltage to ground on the ungrounded conductors is not any greater than 150 volts
- So that the neutral of a 3-phase, 4-wire wye system is used as a circuit conductor
- So that for a 3-phase, 4-wire delta the midpoint of one phase winding is used as a circuit conductor

There are a number of scenarios where you can ground an AC system but you don't have to. They involve systems that range from 50 volts to 1000 volts with ground detectors and that fall into either of the categories below:

- An electrical system that is exclusively used to supply an industrial furnace that is used for melting, refining, or tempering
- Separately derived systems that are only used for rectifiers that supply adjustable-speed industrial drives
- Separately derived systems that are supplied by transformers with a primary voltage that is less than 1000 volts and are only used for control circuits, will only be accessible to qualified persons, have control power continuity, and include ground detectors on the control system

When you are grounding an AC service, you need to install a grounding electrode conductor and connect it to a grounded service conductor at each service. There are a number of circuits listed in **NEC 250.22** that are

---

Trade Tip

A rectifier is a device that converts alternating current to DC or to current with only positive value. This process is referred to as rectification. Rectifiers may be made of solid-state diodes, vacuum-tube diodes, mercury arc valves, and other similar components. A circuit that performs the opposite function of converting DC to AC is an inverter.

Trade Tip

In looking at the list of circuits that you cannot ground, remember not to ground secondary light-system circuits.

not permitted to be grounded. That's right: you are *not* allowed to ground circuits in installations such as health-care-facility anesthetizing locations, induction rooms, or circuits that supply medical equipment that uses 150 volts or more, such as portable X-ray machines. You are also prohibited from grounding circuits for electric cranes that are used over Class III location combustible fibers or for equipment within electrolytic-cell working zones.

Based on **NEC** 250.24, you can install a main bonding jumper at the building service depending on whether an equipment grounding conductor is run with the feeder conductors.

NEC 250.26 lists the specifications for AC-system grounding conductors. These are listed in the table below.

Grounded Conductor verses Grounding Conductor

Let's examine two types of conductors that have such similar names that many electricians use the terms interchangeably or incorrectly. A **grounded conductor** is a system or circuit conductor that is intentionally grounded. For example, neutrals are one type of grounded conductor. Grounded conductors are color-coded with either a white or gray outer

FIGURE 5.1

AC conductor grounding specifications.

NEC 250.26 Requires AC Premise Wiring System Grounded Conductors	
Type of System	**Conductor To Be Used**
Single Phase 2-Wire	One Conductor
Single Phase 3-Wire	The Neutral Conductor
Multiphase with One Wire Common To All Phases	The Common Conductor
Multiphase with One Grounded Phase	One Phase Conductor
Multiphase with One Phase Used as a Single Phase 3-Wire	The Neutral Conductor

Trade Tip

Remember that Article 100 defines a main bonding jumper as a connection between a grounded circuit conductor and the equipment grounding conductor at the electrical service.

finish. The second type of conductor is the **grounding conductor**. This is a conductor that is used to connect equipment or the grounded circuit of a wiring system to a grounding electrode. It describes conductors that attach electrical devices to earth ground; these conductors can be bare, covered, or insulated. The installation order for the two conductors is that the grounded conductor system of a building or structure is connected to the grounding conductor system at the service and only at the service as outlined in **NEC** 250.24(A). The two conductors are connected together by using a main bonding jumper.

Grounding Electrodes

A grounding electrode is any of a number of devices that establishes an electrical connection to the earth. **NEC** 250.52 describes the materials and installation requirements for approved grounding electrodes. The first

---

CODE UPDATE

Section 250.32(B)(2) of the previous code is now incorporated as an exception to Section 250.32(B)(1) and applies to existing installations only. In Section 250.32(B)(2) of the 2005 **NEC** permission to use a grounded conductor for grounding equipment did not include new electrical designs. By changing this requirement to an exception, it forces you to develop designs and installations of feeders or branch circuits that include an equipment grounding conductor in accordance with the requirements of 215.6 and 250.32(B)(1) for any feeders or branch circuits that will be installed to supply separate buildings or structures. In this way inappropriate neutral-to-ground connections are reduced, plus the restrictive conditions of the new exception still have to be met and are subject to approval of the authority having jurisdiction.

method utilizes an underground metal water pipe. The pipe has to be electrically continuous and in direct contact with the ground for at least 10 feet. If the water pipe is interior, it has to be located 5 feet or less from the entrance point to the building, unless it is in an industrial or commercial setting that will only be accessed by a qualified person. The metal frame of a building can also be considered to meet the grounding-electrode requirement if it meets any of the following earth-connection condition: 10 feet or more of one metal structural section must be in direct contact with the earth or encased in concrete that is in direct contact with the earth.

The requirement that all grounding electrodes present at each served building must be bonded together to create a grounding electrode system is the focus of **NEC** 250.50. The Article 100 definition of a grounding electrode is a device that establishes an electrical connection to the ground (earth). There are several vehicles for forming this connection:

- A metal underground pipe
- The metal frame of the building
- A concrete-encased electrode
- A ground ring that encircles the building
- Rod and pipe electrodes
- Plate electrodes

Each of these electrode methods has specific requirements. For example, interior metal water pipes have to be located within 5 feet of the point of entrance to a building in order to be used as grounding electrodes. The metal frame of a building has to have 10 feet or more of structural direct contact to the earth or be encased in concrete that is in direct contact with the earth in order to meet the standard for a grounding electrode. A grounding ring has to encircle the building, be in direct contact with the earth, and consist of at least 20 feet of bare copper conductor that is at least #2 AWG. Pipe or conduit used as electrodes must have a diameter of at least 3/4-inch trade size; iron or steel rods need to be at least 5/8-inch trade size; stainless-steel rods cannot be less than 5/8 inch in diameter; and non-ferrous rods need to be at least 1/2 inch in diameter. There are two types of electrodes that are specifically not allowed to be used as grounding electrodes. These are metal underground gas-pipe systems and aluminum

---

CODE UPDATE

In the 2008 edition of the **NEC**, requirements for a concrete-encased electrode now include vertical electrodes and instructions on what to do when multiple isolated concrete-encased electrodes are present. Structural-steel rebar in vertical foundations is now considered suitable as a grounding electrode as long as it meets all the requirements for horizontal structural-steel rebar electrodes. Additionally, a building with multiple isolated concrete-encased electrodes, such as for spot footings, is only required to use one of these "present" electrodes. Requiring all the concrete-encased electrodes to be bonded together would not provide any more of a safeguard than using just one electrode.

---

electrodes. If more than one of the permissible type of electrode systems are used, than each electrode in the grounding system has to be within 6 feet of the other unless the grounding electrodes are bonded together, in which case they are considered to be a single grounding electrode system.

Let's look at an example of this code's application. Assume that you are going to lay out the grounding system for an industrial piping plant with steel buildings and equipment that is mounted on foundations that are not under cover. The foundations are going to have rebar reinforcement, and the equipment needs to be attached to a grounding loop that has been routed throughout the plant. You need to determine if the foundation rebar is required to be attached to the grounding system or grounded at any point at all. You know from **NEC** 250.52 (A)(1) through (A)(7) that grounding electrode systems that are present at each building or structure that is supplied with electric power are required to be bonded together based on the standards in **NEC** 250.50. If the foundations meet the description in 250.52 (A)(3) for concrete-encased electrodes, then the foundations are required to be part of the grounding electrode system at each building or structure.

Section 250.56 only requires an earth resistance value of 25 ohms for single-rod, pipe, or plate-electrode grounding systems. Additionally, installations have to comply with the overall performance objectives stated in 250.4(A) and (B), which can mean that testing of the grounding electrode system is needed, particularly in the case of connecting a new installation to an existing grounding electrode system that can't be visibly inspected.

You also have to take into consideration that an electrical system must be able to be maintained during its operational life in order to protect people and property from electrical hazards. Keep in mind **NEC** 90.1(B), which outlines that both code compliance and proper maintenance of an installation that is essentially free from hazard may not necessarily result in the most efficient or convenient installation ore one that is adequate for good service or future expansion.

Grounding Electrode Conductors

Table 250.66 in the **NEC** lists the minimum sizes for the grounded electrode conductors of grounded or ungrounded AC systems, as well as for derived conductors of separately derived AC systems. The table deals strictly with the size of the largest ungrounded service-entrance conductor or the equal area for parallel conductors and specifies the minimum size of the grounding electrode conductor.

If multiple sets of service-entrance conductors are involved, then a size equal to the largest service-entrance conductor needs to be determined by calculating the largest sum of the areas of conductors that correspond to each other in each set. However, if no service-entrance conductors are involved, then you have to calculate the largest sum of the areas of the service-entrance conductors that would be needed to service the electrical load involved.

FIGURE 5.2

Minimum size is not based on the application conditions of the grounding electrode.

Minimum Required Size of Grounding Electrode Conductors		Size of the Largest Ungrounded Service-Entrance Conductor/ Equivalent Parallel Conductor Area	
Copper (AWG)	Aluminum or Copper-Clad Aluminum (AWG)	Copper (AWG)	Aluminum or Copper-Clad Aluminum (AWG)
8	6	2 or smaller	1/0 or smaller
6	4	1 or 1/0	2/0 or 3/0
4	2	2/0 or 3/0	4/0 or 250
2	1/0	Over 3/0 but less than 350	Over 250 to 500
1/0	3/0	Over 350 but less than 600	Over 500 to 900
2/0	4/0	Over 600 but less than 1100	Over 900 to 1750
3/0	250	Over 1100	Over 1750

Next, you have to take into consideration the type of grounding electrode application you will be using. For example, if you are connecting a grounding electrode conductor to a rod, pipe, or plate, then the part of the conductor that is the solitary connection to the grounding electrode does not have to be any larger than 6 AWG copper or 4 AWG aluminum wire. For connections to concrete-encased electrodes, the portion of the conductor that is the lone connection to the grounding electrode is not required to be larger than 4 AWG copper wire. If you plan to connect the grounding electrode conductor to a ground ring, then the segment of the conductor that is the sole connection to the grounding electrode does not have to be any larger than the conductor that is used in the grounding ring. **NEC** 250.68 requires that any connections you make to a grounding electrode conductor or from a bonding jumper to a grounding electrode have to not only be accessible but must also create an effective and permanent grounding path.

NEC 250.70 lists the specific means required for connecting grounding electrodes to a grounding or bonding conductor as follows:

- Exothermic welding
- Listed lugs or pressure connectors
- Approved ground clamps that are listed for the grounding conductor materials and rated for direct soil burial or concrete encasement if they are used on pipes, rods, or other buried electrodes

Approved materials for these connections are pipe fittings, pipe plugs, or approved devices that are screwed into the pipe or a pipe fitting; cast bronze, brass, plain, or malleable iron listed bolted clamps; or, for indoor telecommunications installations, rigid sheet-metal straps.

BONDING

Previously we described bonding as joining together metallic parts to form an electrically conductive path. To facilitate this process, you use bonding jumpers. Bonding jumpers are used in several ways and are therefore referred to by different names. The main bonding jumper is installed at the service, and you will only use one. System bonding jumpers are installed at separately derived systems, and you will have one for each separately derived system that is utilized. Both of these types of bonding jumpers do the same thing. They provide an effective ground-fault current path from your equipment enclosures and raceways to the electrical source. Equipment bonding jumpers are installed to tie equipment together in order to keep voltage from building up in equipment and creating a difference in potential between various pieces of equipment. For example, if you were bonding two metal raceways together, you would use an equipment bonding jumper anywhere there was a break in the metal conduit run. A circuit bonding jumper ties together the various conductors of a circuit, and it has to have the ability to carry the same or greater current as the circuit conductors it is connecting.

NEC 250.92 requires that any of the non-current-carrying metal parts of the following types of equipment have to be effectively bonded together:

- Cable trays or cable-bus framework
- Service raceways or cable armor

FIGURE 5.3

A grounded electrode conductor runs through a metal raceway bonded to the conductor with a grounding bushing.

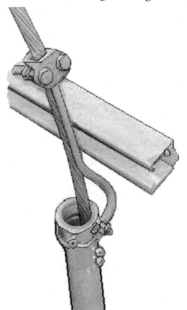

Did You Know?

You have to bond at each end of a raceway as well as to any intervening raceways, boxes, or other enclosures between the service equipment and the grounding electrode.

- Auxiliary gutters
- Service enclosures that house service conductors such as meter fittings or boxes that are inserted in the service armor or raceways
- Any metallic raceway or armor that encloses a grounding electrode conductor

Types of Bonding Jumpers

NEC 250.102 provides standards for equipment bonding jumpers. This type of bonding jumper has to be copper or some other kind of corrosion-resistant material and can be a bus, screw, wire, or similar form of conductor. Equipment bonding jumpers that are installed on the supply side of a

CODE UPDATE

NEC Section 250.94 provisions referring to bonding "other" systems deal with providing an accessible location outside of enclosures for connecting intersystems. This has been reworded in the 2008 edition. The new text adds a requirement for a well-defined, dedicated location for terminating bonding and grounding conductors on a specific set of terminals or a bonding bar. The termination has to have adequate capacity to handle multiple communication systems, such as satellite, CATV, and telecommunications at the premises, and at least three terminals are required. Specifying the termination locations for the intersystem bonding termination is the key part of this revision. The termination means has to be secured electrically and mechanically to the premise meter enclosure that is located at either the service-equipment enclosure or at the grounding electrode conductor. See the figure on the next page for an illustration of this new standard.

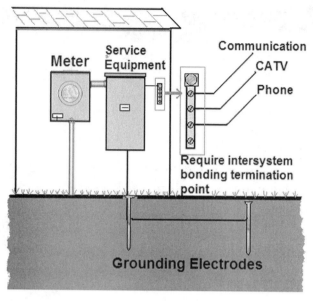

Meter   **Service Equipment**   **Communication**

CATV

Phone

Require intersystem bonding termination point

Grounding Electrodes

FIGURE 5.4

The previous wording of Section 250.94 is now an exception to the new requirements of this standard.

service cannot be any smaller than the sizes for grounding electrode conductors that are listed in **NEC Table 250.66**. Additional requirements are also applied. For example, if the service-entrance phase conductors are bigger than 1100 kcmil copper or 1750 kcmil aluminum, than the equipment bonding jumper has to have an area that is at least 12 1/2 percent of the area size of the largest phase conductor. However, if the phase conductors and the bonding jumper are made of different materials, the minimum size of the bonding jumper needs to be based on the "assumed use" of the phase conductors. This means that you would first assume that the conductors were made of the same materials, even though they are not, and have an equal amperage, and then you would determine the sizing based on that premise. For service-entrance conductors that are installed parallel in two or more raceways, the equipment bonding jumpers must also be run in parallel if they are routed with raceways or cables. In this kind of installation, the bonding-jumper size for each of the raceways or cables would be based on the size of the service-entrance conductors in each of the cables or raceways.

Minimum Copper AWG	Minimum Aluminum or Copper-Clad Aluminum AWG	Amperage Setting/Rating of Automatic Overcurrent Device in the Circuit **Ahead** of the Equipment, Conduit, etc.
14	12	15 amps
12	10	20 amps
10	8	30 amps
10	8	40 amps
10	8	60 amps
8	6	100 amps
6	4	200 amps
4	2	300 amps
3	1	400 amps
2	1/0	500 amps
1	2/0	600 amps
1/0	3/0	800 amps
2/0	4/0	1000 amps
3/0	250	1200 amps
4/0	350	1600 amps
250	400	2000 amps
350	600	2500 amps
400	600	3000 amps
500	800	4000 amps
700	1200	5000 amps
800	1200	6000 amps

FIGURE 5.5

The size of equipment grounding conductors for raceways or equipment must meet these minimum requirements.

Equipment bonding jumpers that are installed on the load side of service overcurrent-protection devices need to have a minimum size that is based on the requirements of **NEC** Table 250.122. However, these bonding jumpers cannot be any smaller than 14 AWG, and they don't have to be any larger than the largest ungrounded circuit conductors that supply the equipment.

Two of the most typical means of bonding are to pipes and metal structures, and bonding jumper methods for these are covered in **NEC** 250.104. Metal water pipes need to be grounded to one of the following:

- The service equipment enclosure
- The grounded conductor at the electrical service
- The grounding electrode conductor if it is sufficient size

According to Section 250.104(A) you need to refer to Table 250.66 to determine the required sizes for these bonding jumpers. For example, if you needed to bond the interior metal water piping of a building with an electric service that is supplied with 3/0 copper, you would use a 4 AWG copper bonding jumper. However, **NEC** 250.104(A) describes multiple-occupancy-building bonding requirements. Let's look at an example of how to apply these requirements.

Assume that a service is supplied with two parallel sets of 500 kcmil; then, based on this Section and its reference to Table 250.66, it would require a 2/0 AWG copper bonding jumper. Now you might ask yourself, if the cold water is bonded at the water service, why does the hot-water piping require such a large bonding jumper? First, look at the general requirement for bonding metal water piping systems that is contained in 250.104(A). This states that any metal piping system used for water, whether it is for a domestic supply or other use, is treated as a metal water piping system and is required to be bonded to the electrical services and systems that are in or on the same building or area of a building. If it is determined that the piping system is electrically continuous, a single connection to any point on the piping system will comply with this requirement. The bonding conductor or jumper is sized using Table 250.66. Next, if there is some dielectric separating the hot-water from the cold-water piping, a bonding jumper of the same size as the bonding conductor run from the service equipment or separately derived system is required. This ensures that all of the metal water piping system is bonded using a fully sized bonding conductor or jumper. By looking back at Table 250.66 to pick the bonding conductors and bonding jumpers, you will see that this standard is providing a worst-case-scenario approach. This ensures that there is a sufficient path for ground-fault current in the event that the metal water piping becomes energized because the electrical system fails somewhere. In addition to supplying this ground-fault current path, bonding of the metal water piping system serves as an equipotential connection to reduce the likelihood of electric-shock hazards at interfaces where the conductive surfaces of equipment supplied by the electrical system and by the water system are both present.

Equipment Grounding

Equipment that is fastened in place or connected with fixed, permanent wiring methods must have the exposed non-current-carrying metal parts grounded based on **NEC** 250.110. The types of equipment that have to be grounded regardless of the voltage involved are listed below:

- Switchboard frames and structures, unless it is a DC 2-wire switchboard that is already effectively insulated from ground
- Pipe organs
- Motor frames and motor controller enclosures
- Elevators and cranes
- Commercial garages, theaters, and motion-picture-studio electrical equipment, except for pendant lampholders with circuits of 150 volts to ground or less; projection equipment must also be grounded
- Electric signs
- Power-limited remote-control signaling and fire-alarm circuits that have system grounding requirements provided in **NEC** 250.112 Part II or Part IV
- Light fixtures
- Permanently mounted skid equipment
- Motor-operated water pumps
- Metal well casings

There are fourteen types of equipment grounding conductors listed in **NEC** 250.118 that are approved for use to enclose the conductors of power-supply circuits or to run with them. These can be used individually or in combination and are as follows:

- Solid or stranded copper, aluminum, or copper-clad aluminum conductors, which may be insulated, covered, or bare and can be in the form of a wire or busbar of any shape
- Rigid metal conduit
- Intermediate metal conduit
- Electrical metal tubing
- Flexible metal conduits that meet one of the following criteria: are terminated in fittings that are approved for grounding; are protected by 20 amp or less OCPDs; have a combined length of

flexible conduit materials in the same ground return path that does not exceed 6 feet; will have an equipment grounding conductor installed

- Flexible metallic tubing that is listed for grounding and has conductors in conduit that are protected by 20 amp or less OCPDs as well as a combined length of flexible and liquid-tight conduit in the same ground return path that does not exceed 6 feet

- Listed liquid-tight flexible metal conduit that meets the following standards:

  a) are terminated in fittings that are approved for grounding; are 3/8 to 1/2 inch (#12 to #16) and protected by 20 amp or less OCPDs.

  b) are 3/4 to 1 1/4-inch (#21 to #35) and protected by OCPDs th rated at 60 amps or less and not run with flexible metal or liquid-tight conduit 3/8" to 1/2-inch (#12 to #16) in the grounding path.

  c) do not exceed 6 feet in the total combined length of the flexible and liquid-tight conduits in the same ground path; have an equipment grounding conductor is installed.

- Copper sheath of metal-sheathed, mineral-insulated cable

- Type AC cable armor

- Type MC cable that is identified and listed for grounding, as long as the metal sheath and grounding conductor are of interlocked metal tape or type MC cable or are smooth or corrugated tube-type MC cable

- Cable trays

- Cable-bus frameworks

- Surface metal raceways that are listed for grounding

- Other listed metal raceways and auxiliary gutters

- Cable assemblies and flexible cords as outlined in **NEC** 250.138

The standards for sizing equipment grounding conductors are provided in **NEC** 250.122. For the general rules for sizing, refer to **NEC** Table 250.122.

> **Trade Tip**
>
> Remember that if you have to increase the size of ungrounded conductors, you need to bump up the size of the equipment grounding conductors proportionally to the circular mil area of the ungrounded conductors.

For multiple circuits that are run with a single equipment grounding conductor in the same raceway, you need to be sure to size the equipment grounding conductor for the largest overcurrent device that protects the raceway conductors. Motor circuits with overcurrent devices that consist of an instantaneous trip circuit or a motor short-circuit protector can have an equipment grounding conductor that is sized based on the rating of the motor overload-protection device as long as you don't use a size that is less than that listed in **NEC** Table 250.122.

Equipment grounding conductors in a flexible cord that has a 10 AWG or smaller circuit conductor as the largest size cannot be smaller than 18 AWG copper, and they cannot be any smaller than the circuit conductor size. If the circuit-conductor size is more than 10 AWG, then you have to refer to **NEC** Table 205.122 for the required size. If the equipment in a system has ground-fault protection installed, each parallel equipment grounding conductor that is in a multiconductor cable should be sized from Table 250.112 based on the trip rating of the ground-fault protection. Additionally, the ground-fault protection needs to be installed in a manner so that only qualified people can maintain or service the installation. The ground-fault protection equipment must also be set to trip at no more than the amperage of any single ungrounded conductor of one of the cables in parallel, and it has to be rated for use in protecting the equipment grounding conductor.

Grounded Neutral Conductors

Keeping the normal current that flows across a neutral conductor on the path it is intended to follow is an important aspect of safety and noise control over the grounded paths, including the equipment grounding conductors of a system. Section 250.142(B), Exceptions 1 through 4, provide restrictive conditions for using the grounded (neutral) conductor for grounding on the load side of the service disconnect as follows:

- On the supply-side of equipment you can ground non-current-carrying metal parts of equipment, raceways, and other enclosures with a grounded circuit conductor only at these locations:

  - On the supply side or within the enclosure of the AC service-disconnecting means

  - On the supply side, or within the enclosure of, the main disconnecting means for separate buildings (Refer to **NEC** 250.32(B) as a cross-reference for the standards for separate buildings, which require that the equipment grounding conductors be run with supply conductors and connected to the building structures or disconnecting means and to the grounding electrodes.)

  - On the supply side or within the enclosure of the main disconnecting means or overcurrent devices of a separately derived system as permitted by **NEC** 250.30(A)(1)

- On the load side of equipment of the service disconnecting means, the load side of a separately derived system's disconnecting means, or the overcurrent devices for a separately derived system that does not have a main disconnecting means, you can only use a grounded circuit conductor for grounding the non-current-carrying metal parts of equipment as permitted in 250.30(A)(1) and 250.32(B)

Grounding High-Voltage Systems

When grounding high-voltage systems that are 1kV or higher, you need to follow the standards in **NEC** Part X. You can use a system neutral that is derived from a grounding transformer to ground high-voltage systems. A single-point grounded or multigrounded neutral can be used for solidly grounded neutral systems; the neutral conductor needs to have a minimum insulation level of 600 volts. The exceptions would be for the neutral conductors in overhead installations, service entrances, and direct buried portions of feeders, in which case the neutral conductors can be bare copper.

Trade Tip

NEC 250.30(A)(1) is very detailed and includes a number of exceptions for grounding the electrode conductor in multiple separately derived systems. It begins by allowing the grounded neutral terminal of each derived system to be grounded to a common grounding electrode conductor. The grounding electrode conductor and grounding electrode tap must comply with the requirements listed in (a) through (c); however, there are three exceptions and three other provisions:

- If the system bonding jumper is a wire or bus bar, then the grounding electrode tap can terminate at the equipment grounding terminal, bar, or bus on the metal enclosure of the separately derived system.
- Separately derived systems that are rated 1kVA or less are not required to be grounded, but to ensure that ground faults can be cleared, a system bonding jumper must be installed.
- The system bonding jumper cannot be smaller than 14 AWG copper or 12 AWG aluminum for a system that supplies a Class 1 through Class 3 circuit derived from a transformer that is rated no larger than 1000 volt-amperes. Additionally, it cannot be smaller than the derived phase conductors.
- The common grounding electrode conductor cannot be smaller than 3/0 AWG copper or 250kcmil aluminum.
- Each grounding electrode tap needs to be sized in accordance with 250.66, based on the largest separately derived ungrounded conductor of the separately derived system.
- All grounding electrode tap connections must be made at an accessible location by either listed connections to aluminum or copper bus bars or exothermic welds. The bus bar cannot smaller than 1/4 inch by 2 inches. If aluminum bus bars are used, the installation must also comply with 250.64(A). Grounding electrode tap conductors must be connected to the common grounding electrode conductor so that the latter isn't spliced.

The ampacity of the high-voltage neutral conductor has to be sufficient to carry the load and not less than 33 1/3 percent of the phase conductor amperage unless it is for an industrial or commercial installation. For industrial and commercial premises, you can size the neutral conductor at no less than 20 percent of the phase-conductor amperage as long as it is done

CODE UPDATE

In the 2008 edition of the **NEC**, a new sentence was added to **NEC** 250.146: An equipment bonding jumper is not required for receptacles attached to listed exposed work covers when conditions are met. These conditions are that the receptacle is attached to the cover with two permanent rivets or has a threaded or screwed locking means and that the cover mounting holes are located on a flat nonraised portion of the cover. This is because exposed work covers with two fasteners used to attach the receptacle to the cover are considered a suitable bonding means.

with engineering supervision. If you intend to use a single-point grounded system, you need to refer to **NEC** 250.184(B) 1 through 8 for the standard requirements. Portable and mobile high-voltage equipment must be supplied from a system that has its neutral grounded through an impedance. If you are connecting to a delta-connected high-voltage system, a system neutral needs to be derived. Finally, the voltage that is developed between the portable or mobile equipment and the ground by the

FIGURE 5.6

Solidly grounded system.

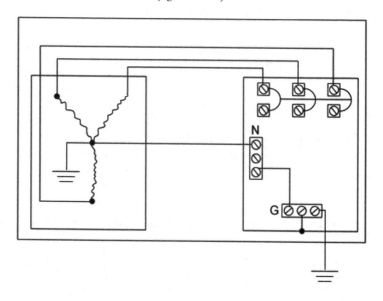

flow of the maximum ground-fault current cannot be greater than 100 volts. Grounding conductors that are not an integral part of a cable assembly cannot be smaller than 6 AWG copper or 4 AWG aluminum, and the metal, non-current-carrying parts of the mobile or portable equipment, enclosures, or fencing must be grounded.

Next, we will cover the methods and materials for wiring that are provided in Chapter 3 of the **NEC**. You will often need to refer back to subjects we have covered previously.

CHAPTER

6

Wiring Methods

WHAT YOU NEED TO KNOW

Conductors are used to carry electrical current from the power source, such as the utility feed, to the point of use, such as a piece of equipment or an outlet. There are three basic properties of all conductors: the type of material, such as copper or aluminum; the insulation type; and the conductor size. Conductors have to be run in cable assemblies or raceways. The materials, structure, and installation standards for cable and conduit systems are referred to as wiring methods. **NEC** Article 300, Part 1, covers all of the wiring requirements for systems that operate at 600 volts or less, and Part II outlines systems over 600 volts.

TERMS TO KNOW

Many of these terms are found over and over in the **NEC**. For example, grounding conductors and equipment grounding conductors are likely to be mentioned in a number of standards. Here are several of the terms that apply to wiring methods and materials.

Ampacity: Also known as current rating, this is the RMS electric current that a device can carry without damage within specified temperature limitations in a particular environment, depending on temperature ratings, potential power loss, and heat dissipation.

AWG: Acronym for American Wire Gauge, this is the American standard established for nonferrous wire conductor sizes. The thicker the wire, the greater its current-carrying capacity and the longer the distance it can span. The gauge indicates the diameter. The smaller the AWG number, the thicker the wire. Metal is pulled through a series of increasingly smaller dies to create the final wire size, and the AWG number is the number of dies. The more dies, the larger the number and the smaller the diameter.

Cable: An assembly of several conductors that are bound together and usually jacketed by a metallic or nonmetallic material. Cable assemblies come in a wide variety of sizes, types, and coverings.

Conductor of the Same Circuit: Conductors that are run to the same circuit must be run together in the same raceway, cable tray, or auxiliary gutter. There are some exceptions to this requirement, which are listed in **NEC** 300.3(B) 1 through 4.

Enclosure: An auxiliary gutter that runs between a column-width panelboard and a pull box that includes neutral terminations. The neutral conductors of circuits supplied from the panelboard are permitted to originate in the pull box.

FIGURE 6.1

#40 wire is smaller than #18.

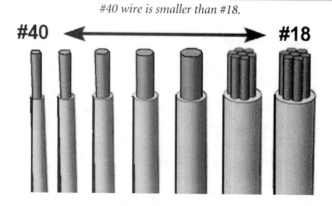

#40 ⟵⟶ #18

Grounding Conductor: A conductor used to connect equipment or the grounded circuit of a wiring system to a grounding electrode or electrodes.

Equipment: A general term, including material, fittings, devices, appliances, luminaires, apparatus, machinery, and the like, used as a part of or in connection with an electrical installation.

Nonferrous Wire: A metal that does not contain iron. Aluminum and copper are common nonferrous metals.

Parallel Installation: Copper, aluminum, and copper-clad aluminum conductors that are at least #10 AWG and make up each phase' polarity, neutral or grounded circuit conductors can be electrically joined at both ends (connected in parallel).

Plenum. A compartment or chamber that connects one or more air ducts and forms part of an air distribution system.

Premises Wiring: Interior and exterior wiring, including power, lighting, control and signal circuit wiring, and all their associated hardware, fittings, and wiring devices. Premise systems include the wiring from the service point or power source to the outlets or the wiring from the power source to outlets with no service point. Premises wiring does not include the inside of appliances, light fixtures, motors, controllers, or motor control centers.

Raceway. An enclosed channel of metal or nonmetallic materials that is designed specifically for holding wires and cables. Raceways include rigid metal conduit, rigid nonmetallic conduit, intermediate metal conduit, liquid-tight flexible conduit, flexible metallic tubing, flexible metal conduit, electrical nonmetallic tubing, electrical metallic tubing, underfloor raceways, cellular concrete floor raceways, cellular metal floor raceways, surface raceways, wireways, and busways.

Single Conductor: An individual conductor only installed under the conditions and standards provided in **NEC** Chapter 3.

OVERVIEW OF WIRING METHODS

Single-conductor building wire is generally a small, solid wire, because the wiring does not have to be very flexible. Cables that will be used for very flexible installations or in marine applications may be protected by woven

Trade Tip

Single conductors can only be used where they are part of a standard wiring method. "Standard" means a wiring method that is covered specifically by the **NEC**.

bronze wires. When you use building wire conductors that are larger than 10 AWG, the wires are stranded for flexibility during installation. Industrial power and control cables may contain numerous insulated jacketed conductors with helical tape steel or aluminum armor, steel wire armor, or even an overall PVC or lead jacket for protection from moisture and physical damage. Power and communications cables for applications such as computer networking that are routed in or through plenums in office buildings are required to either be encased in metal conduit or rated for low flame and smoke production.

Insulated cables are rated based on their permitted operating voltage and their maximum approved operating temperature at the conductor surface. One cable can have several usage ratings—for example, one rating for dry installations and another for exposure to moisture. For some industrial uses, such as in steel mills and similar hot environments, you cannot use organic conductor insulation materials. In these situations, cables insulated with compressed mica flakes are sometimes used. Another example of a high-temperature cable is mineral-insulated cable that has individual conductors within a copper tube; the space is filled with magnesium oxide powder. The whole assembly is drawn down to a smaller size, which compresses the powder. Such cables have a certified fire-resistance rating and are less flexible. Rubberlike synthetic polymer insulation is used in industrial cables and power cables that are installed underground because it provides better moisture resistance.

Multiple conductors that are bundled together in a cable cannot dissipate heat as easily as single insulated conductors, so these multiconductors are always rated at a lower ampacity. Throughout the **NEC** there are standard tables that cross-reference approved conductors based on the maximum allowable current for a particular size of conductor, for the voltage and

---

CODE UPDATE

A new Section 300.9 in the 2008 publication of the **NEC** is titled "Raceways in Wet Locations Above Grade." It explains that when raceways are installed in wet locations above grade, the interior of the raceways are be considered to be a wet location. Insulated conductors and cables installed in raceways in wet locations above grade must comply with 310.8(C).

---

temperature rating at the surface of the conductor for specific physical environments, and for the required insulation type and thickness. If a cable runs through several difference environmental areas, the most severe condition determines the required rating of the overall conductor run.

Where cables enter electrical devices, they need to be secured by approved fittings such as screw clamps for jacketed cables in a dry location. In a wet location, the fitting might be a polymer-gasketed cable connector that crimps the armor of an armored cable and provides a water-resistant connection. Special cable fittings are required to reduce the risk of explosive gases affecting the interior of jacketed cables if the cable passes through a location where inflammable gases are present.

To prevent individual cable-conductor connections from loosening, cables have to be supported near where they enter devices and at regular intervals throughout the cable run. There are specific requirements for how far apart cable supports can be and the methods that are allowed for securing the supports to various types of materials such as drywall, wood, and concrete. In tall buildings special designs are required to support conductors in vertical cable runs, and usually only one cable is allowed per fitting. **NEC** Table 300.19 (A) lists various sizes and types of wire and the required spacing for conductor supports.

Protecting Wire from Damage

Beginning with **NEC** Article 300, wiring methods and ways to protect wiring from damage are outlined. **NEC** Table 300.5 provides the minimum burial depth for various types of installations between 0 volts and 600 volts,

---

Fast Fact

Because aluminum is nonmagnetic, it eliminates heating due to hysteresis, which is the lag time between when a change is initiated and when the effect of that change takes place. Magnetic materials hold onto energy even after current has passed through them. Although aluminum does not produce heat, aluminum conduit, locknuts, and enclosures do carry eddy currents.

---

nominal. For example, rigid metal conduit that is installed under the driveway of a single-family home has to be buried at least 18 inches underground, but if the same rigid metal conduit is installed under a street or highway, it must be at least 24 inches deep.

As provided in **NEC** 300.3(B), you cannot run two conductors of a circuit in one cable and another conductor from the same circuit in a separate cable or raceway. All ungrounded conductors, neutral conductors, and equipment grounding conductors of a single circuit have to be run in the same raceway or cable to prevent heat induction that could result in damage to the conductors, fire, or reduced impedance if a ground fault developed.

When separate conductors are in a nonmetallic raceway, as permitted in 300.5(I) Ex. 2, you can minimize inductive heat to the enclosure by using aluminum locknuts or cutting a slot between the individual holes that the conductors pass through.

Power conductors of different systems can occupy the same raceway, cable, or enclosure if all conductors have an insulation voltage rating that is not less than the maximum circuit. On the other hand, you do have to separate control, signal, and communications wiring from power and lighting circuits so that the higher-voltage conductors do not accidentally energize them. Exceptions to this requirement allow power conductors to terminate at listed signaling equipment if the power conductors are separated by at least 1/4 inch from the low-voltage conductors. All conductors have to be protected from physical damage using the methods specified in **NEC** 300.4, which include the following requirements:

- Parallel conductors may be used, but each raceway must contain all the conductors of the same circuit.

- **NEC** 300.4(A) requires that cables or raceways that are installed through bored holes in wood must be at least 1 1/4 inches away from the edge of the wood. If this clearance cannot be met, then you need to install a 1/16-inch steel plate to cover the area of the wiring. You can notch the wood as long as the strength of the structure won't be jeopardized. Wiring can be laid in the notches, but a 1/16-inch steel plate has to be installed to cover the area. A steel plate is not required if you are installing rigid metal conduit, intermediate metal conduit, electrical metallic tubing, or rigid nonmetallic conduit.

- **NEC** 300.4(B) explains that if you are installing nonmetallic-sheathed cables through metal framing, such as metal studs, a listed grommet or bushing has to be installed to protect the hole that the cable passes through. In locations where nails or screws could damage nonmetallic-sheathed cables or electrical nonmetallic tubing, a 1/16-inch steel plate needs to be installed to cover the area.

- Direct buried cables have to be rated for underground installation and at the minimum depth requirements listed in **NEC** Table 300.5. Cables installed under a building must be in a raceway that extends the full length of the building and beyond the outside wall. Direct buried conductors and cables have to be protected by an enclosure or raceway from below where they emerge from the ground to a point 8 feet above the grade level, per **NEC** 300.5(D)(1). Underground conductors that enter a building must be protected up to the point where they enter, based on **NEC** 300.5(D)(2). Warning tape has to be run for service conductors that are not encased in concrete and are buried 18 inches or more below grade. The warning tape needs to be run in a trench that is at least 12 inches above the underground service installation, per **NEC** 300.5(D)(3). Directly buried conductors can be spliced using approved methods without a junction box. Cables that terminate underground and have protruding directly buried wiring must be terminated with a bushing or similar sealing compound. All conductors for the same circuit need to be installed in the same raceway, or, if you are using open conductors or cables, close to each other in the trench. When you backfill underground installations, you have to use caution not to damage the conductors. A layer of sand or gravel can be placed over the conductors to protect them, or you can use running boards or sleeves as a means of protection, as explained in **NEC** 300.5.

- Conduits or raceways that could allow moisture to contact live parts must be sealed. Any metal raceways, cables, or fittings used must be suitable for the area in which they are installed. If corrosive conditions exist, then raceways have to be covered with a corrosion-resistant coating or other suitable protection. In wet locations, including uses where the walls are frequently washed such as dairies, all the raceways, cables, and enclosures must be installed with at least 1/4 inch of air space between them and the surface that they are mounted to. This requirement also applies when raceways, cables, or enclosures are mounted on absorbent materials such as damp wood. The insides of raceways in wet locations that are located above ground are considered wet locations, and all the conductor types used inside these raceways must comply with **NEC** 310.8(C).

Electrical Boxes

Boxes or fittings are required at every splice or pull point in raceways, as well as at every outlet, switch, junction, and termination point, for AC, MC, MI, and NM cables, as listed in **NEC** 300.15. However, a box is not required for splices or taps in direct-buried cables and conductors. Boxes are also not required where a splice box, switch, terminal, or pull point is located in a cutout box or cabinet, in a motor control center, or in an enclosure for a switch.

Trade Tip

Although warning tape is not required for branch and feeder circuits, running it in the trench above these conductors is an inexpensive and easy way to prevent accidental damage to the conductors.

CODE UPDATE

The 2008 edition of the **NEC**, Section 300.5(B), clarifies the issue of exactly what constitutes a wet space by stating explicitly that the *insides* of underground raceways are considered wet locations.

Fittings and connectors can only be used with the type of system that they were designed for. An example would be that you cannot use flexible metal conduit where rigid conduit is required, as specified in **NEC** 300.16. You need to use a box or fitting with a separately bushed hole for each conductor if a change is made between raceway or cable systems and open or knob-and-tube wiring methods. If a raceway terminates behind a switchboard, then you can use a bushing on the end of the raceway instead of a box with bushed holes. You have to make sure that the size and quantity of the conductors you install in a raceway will be adequate to dissipate heat and that they can be installed or removed without damaging the conductors or their insulation.

Raceways

NEC 300.18 provides the standards for raceway installations. Raceways must be completely installed before conductors are pulled into the raceway system. The requirement is that raceways have to be installed complete between outlet, junction, or splice points before conductors are installed. At least 6 inches of free conductor must be left at the end of every box, fitting, or other splice point, per **NEC** 300.14. Conductors that are fed through, which means not spliced in a box, are exempt from this requirement.

Conductors that are installed in vertical raceways must be supported. The requirements for support are provided in **NEC** Table 300.19(A). You are required to use raceways, fittings, and supports that are suitable for

Trade Tip

Even though it is not required, it is good practice to leave extra lengths of feed-through free conductors, because you never know what modifications may be needed in the future.

Trade Tip

If condensation from temperature changes might occur in an installation, then fill the raceway with a material approved by the AHJ that will prevent the circulation of warm air to a colder section of the raceway.

the environment in which you are installing them. For example, if corrosion protection is necessary for underground or wet locations and the conduit is threaded in the field, then you need to coat the threads with an approved electrically conductive, corrosion-resistant compound.

If the vertical rise of a raceway exceeds the values of Table 300.19(A), you must support the conductors at the top or as close to the top as practical (Figure 300-48). The weight of long vertical runs of conductors can cause the conductors to drop out of the raceway if you don't secure them properly. There have been cases where conductors in a vertical raceway were released from the pulling basket (at the top), and the conductors fell down and out of the raceway, injuring the workers!

Material listed for the specific types of wiring methods and construction structures are located in **NEC** 300.21.

FIGURE 6.2

A raceway.

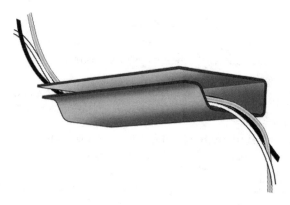

In general, you must keep wiring and air handling separate (Figure 300-52). You can wire in environmental air space under certain conditions listed in 300.21(C).

You must install electrical circuits and equipment in a way that doesn't substantially increase the possible spread of fire or products of combustion. Circuits over 600 volts and circuits 600 volts or less cannot be in the same raceway, cable, or enclosure, with the exception of light-fixture ballast conductors and control instrument conductors. A space not used for environmental-air-handling purposes has no restrictions on wiring methods, so you can use nonplenum cables for those installations.

A number of requirements that pertain to raceways are found in Chapter 3 of the **NEC** and include:

- Metal raceways must be properly grounded. Raceway grounding standards are found in **NEC** 240.4 and 240.5 and Section 300.10; they require that metal raceways and enclosures have to be electrically continuous.

- You must prevent air flow through a raceway by sealing if the two ends of a run are exposed to significantly different temperatures (**NEC** 300.7).

- Raceways or cable trays that contain electric wiring cannot be run with any other type of piping, such as steam, gas, air, or drainage (**NEC** 300.8).

- All raceways, cables, and boxes, except for nonmetallic boxes, have to be electrically and mechanically joined together (**NEC** 300.10 and 300.12).

- You can support electrical wiring to independent support wires within the area of a floor-ceiling or roof-ceiling assembly if they are secured at each end (**NEC** 300.11(A)1). Suspended ceiling support wires can only be used to support wiring for fire-rated systems and only if they have been tested as part of the system or are installed per the manufacturer's instructions for non-fire-rated systems (**NEC** 300.11(A)(2). For these installation, the support wires must be easily distinguished from the suspended ceiling support wires by color, tagging, or other effective means, unless the area is non-fire-rated (**NEC** 300.11(A)2).

- In areas that could cause thermal expansion and contraction, you have to install raceways with expansion fittings. **NEC** Table 352.44(A) provides the expansion characteristics for PVC rigid nonmetallic conduit. In order to determine the expansion characteristics for metal raceways such as EMT, IMC, and RMC, you have to multiply the values in Table 352.44(A) by 0.20.

- If you have to use expansion fittings for a metal raceway, you must also use a bonding jumper to maintain the equipment grounding path that is required by **NEC** 250.98 and 300.10.

- Join all metal raceways, cable, boxes, and fittings to form a continuous low-impedance ground-fault current path. The ground-fault current path has to be adequate to carry any fault that could be imposed on it (**NEC** 110.10, 250.4(A)(3), and 250.22).

High-Voltage Wiring Requirements

For safety reasons, covers have to be installed on all boxes, fittings, and similar enclosures to prevent accidental contact with energized parts, per **NEC** 300.31. AC or DC circuits that are 600 volts or less are allowed to share a raceway, cable, or enclosure, as long as the insulation for each conductor is equal to the highest voltage present. Therefore, a 120-volt circuit can occupy the same raceway as a 600-volt circuit as long as every conductor in the raceway is insulated for 600 volts. Above-ground conductors must be installed in rigid metal conduit, intermediate metal conduit, electrical metallic tubing, rigid nonmetallic conduit, or cable trays. The bending radius requirement of **NEC** 300.34 is now 12 times the *overall* diameter of the shielded or lead-covered conductor.

Did You Know?

Raceways cannot be used to support other raceways, cables, or nonelectric equipment, unless they are Class 2 or Class 3 cables that are only used for equipment control circuits. These cables cannot be installed in the same raceway as the power conductors, per **NEC** 725.55(A), so the best way to attach them is to a raceway.

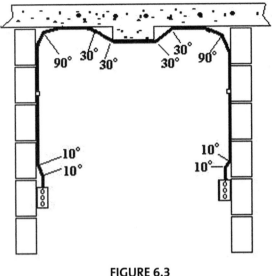

FIGURE 6.3

Bend radius.

As with lower-voltage conductors, high-voltage cables have to be protected against moisture and damage. Minimum ground-cover requirements are provided in **NEC** Table 300.50. Additionally, you must backfill underground installations carefully to ensure that raceways will not be damaged. Shielded and nonshielded underground cables have to be grounded through an effective grounding path. Wherever a raceway enters a building from an underground system, the connection has to be sealed to protect the conductors from moisture or gases.

Now that we have reviewed wiring methods, we will take a detailed look in the next chapter at conductors.

7

Conductors

WHAT YOU NEED TO KNOW

Article 300 is short on words and long on examples provided in tables. When working with conductors, there are a number of critical elements to consider. First, there is the conductor's insulation material and value. For the most part, all conductors are insulated and their use is governed by maximum operating temperatures and ambient installation temperatures. The types of insulation and the conductor material determine the appropriate use and application of each kind of conductor. This leads us to the next important consideration when working with conductors, which is the environment in which the conductors will be used—wet, dry, exposed, or corrosive. Other factors are the voltage the conductor can carry and the number of conductors that are run together. These considerations dictate whether or not any correction factors are required for the conductor installation. Article 310 of the **NEC** is full of tables that combine these various characteristics and list the permitted uses and applications of the many types of approved conductors.

TERMS TO KNOW

Ambient Temperature: The normal air-temperature range in the area that surrounds an electrical conductor.

Circular mil: the standard unit of measure of a wire's cross-sectional area, which is expressed in mils squared

Conductor: A material that can carry electrical current. Voltage applied across a conductor creates an electric current. Conductors referenced in **NEC** Article 310 are made of aluminum, copper, or copper-clad aluminum.

Conductor Ampacity: The total ampere load that the **NEC** allows a conductor to carry in order to meet minimum safety standards after any necessary deration calculations are performed.

Conductor Ampacity Correction Factors: The maximum ampere load maximum permitted in **NEC** Table 310.16 multiplied by ambient temperature correction factors. The **NEC** requires that you reduce the ampacity of any conductor due to ambient temperature listed at the bottom of the table. Then you have to factor in any ampacity deration based on **NEC** Article 310.15.B.2.A.

Correction Factor: Any variable condition that must be accounted for when selecting a wire size. Correction factors such as temperature, number of conductors, and conductor length influence allowable ampacity.

Current-Carrying Conductors: Any conductor found in a conduit that is not a true neutral. Current-carrying conductors cannot be bare wires or green equipment grounding conductors.

Deration: A reduction in the ampacity of a conductor due to correction factors. Conductors are rated for a specific set of conditions, and when those conditions change, ampacity must be derationed.

Dielectric: An insulating material that separates conductors from their outer protective covering. Dielectrics are nonmetallic.

Resistance: Electrical opposition to current flow. Resistance is measured in ohms.

Temperature Rating: Also referred to as the "operating temperature," this is the maximum temperature that a conductor can withstand over a prolonged period of time without serious degradation at any location along its length. Factors that affect a conductor's operating temperature are the surrounding ambient temperature, internal heat that is generated by the conductor carrying the load from its current flow, heat dissipation, and adjacent load-carrying conductors, which can raise the ambient temperature and impede heat dissipation just by being in close proximity to other conductors.

Thermal Resistivity: The capability of a substance to transfer heat through conduction. It is designated "Rho" and expressed in units as degrees C—cm/watts.

Voltage Drop: The reduction in voltage levels from the source to the load, which is caused by conductor resistance.

CONDUCTOR BASICS

Article 310 of the **NEC** starts off with some basic standards for conductors, such as the requirement that they be insulated, stranded if they are over 8 AWG, and installed in raceways. They may be run in parallel as long as they are over 1/0 AWG and comprise each phase, polarity, neutral, or grounded circuit conductor. The code next considers the minimum size requirements for conductors, based on their voltage rating as illustrated below.

FIGURE 7.1

Minimum conductor sizes permitted by NEC 310.5.

MINIMUM CONDUCTOR SIZES		
CONDUCTOR VOLTAGE	**Copper**	**Aluminum or Copper-Clad Aluminum**
0 - 2000 VOLTS	#14 AWG	#12 AWG
2100 - 4000 VOLTS 4001 - 8000 VOLTS	#8 AWG	#8 AWG
8001 - 10,000 VOLTS 10,001 - 15,000 VOLTS	#2 AWG	#2 AWG
15,001 - 25,000 VOLTS 25,001 - 28,000 VOLTS	#1 AWG	#1 AWG
28,001 - 35,000 VOLTS	1/0 AWG	1/0 AWG

Solid dielectric insulated conductors rated above 2000 volts must have ozone-resistant insulation and be shielded, with a few exceptions. Direct-buried conductors must be identified and rated for underground use.

NEC 310.8 discusses various types of installation locations and the conductors approved for each use. Before we look at what kind of conductors can be installed, let's define the three different kinds of locations:

- **Dry locations** are areas that are not normally subjected to moisture, such as inside the framing of a single-family home. Conductors classified for dry locations may be temporarily subject to dampness, as in the case of a house that is under construction.

- **Damp locations** are those that are protected from weather and not subject to being saturated with water or other liquids but that are exposed to moderate degrees of moisture. Examples of these locations include electrical installations that are partially protected from the elements under canopies or overhangs or interior locations such as basements or cold storage buildings.

- **Wet locations** are installations underground or in concrete slabs or masonry that are in direct contact with the earth, are subjected to saturation by water or other liquids, such as vehicle-washing bays or docks, or are in unprotected areas exposed to weather.

Some conductor types can be used in all or several of these locations, while others cannot. See the listing below.

FIGURE 7.2

Conductors such as MTW and RHW can be used in wet, damp or dry locations.

CONDUCTOR LOCATION	CONDUCTOR TYPE
Dry and Damp Locations	*FEB, FEPB*, **MTW**, *PFA, RHH*, **RHW, RHW-2**, *SA, THHN*, **THW, THW-2, THHW, THHW-2, THWN, THWN-2**, **TW**, *XHH*, **XHHW, XHHW-2**, *Z,* **ZW**
Wet Locations Only	MTW, RHW, RHW-2, THW, THW-2, THHW, THHW-2, THWN, THWN-2, TW, XHHW, XHHW-2, ZW

Conductors must be marked, as required in **NEC** 310.11, with the maximum voltage rating, type of wire designation, manufacturer's name or product identifier, and AWG size or circular mil area. An explanation of conductor-type names, approved application locations, maximum operating temperatures, insulation types, and AWG sizes are found in **NEC** Table 310.13. You must use this table to select the cable type(s) appropriate for your particular installation. The information on circular mils allows you to calculate wire fill, which is a big consideration when you are laying out an installation.

TYPES OF CONDUCTORS

While **NEC** Table 310.13 gives a brief description, based on trade names, of various conductors, we'll expound on some if the types of conductors covered in this part of the code:

- **CT** is a designation that is given to cables that meet established UL requirements for cable-tray use.
- **FEP** signifies a fluorinated-ethylene-propylene insulation and jacket compound.
- **MC** stands for metal-clad cable. These conductors are used as power and control cables and are enclosed in rather a smooth (Okoclad®) or a welded and corrugated metallic sheath.
- **RHH** is a designation for conductors with heat-resistant rubber or XLPE insulation that are designed for use in dry locations.
- **RHW-2** designates a conductor that has heat- and moisture-resistant rubber or XLPE insulation and is used in 90-degree C wet or dry locations.
- **THHN** represents PVC-insulated nylon-jacketed conductors for use in dry locations.

FIGURE 7.3

NEC Table 310.13.

Table 310.13(A) Conductor Applications and Insulations Rated 600 Volts

Trade Name	Type Letter	Maximum Operating Temperature	Application Provisions	Insulation	Thickness of Insulation			Outer Covering
					AWG or kcmil	mm	mils	
Fluorinated ethylene propylene	FEP or FEPB	90°C 194°F	Dry and damp locations	Fluorinated ethylene propylene	14–10 8–2	0.51 0.76	20 30	None
		200°C 392°F	Dry locations — special applications[2]	Fluorinated ethylene propylene	14–8	0.36	14	Glass braid
					6–2	0.36	14	Glass or other suit braid material
Mineral insulation (metal sheathed)	MI	90°C 194°F	Dry and wet locations	Magnesium oxide	18–16[3] 16–10	0.58 0.91	23 36	Copper or alloy ste
		250°C 482°F	For special applications[2]		9–4 3–500	1.27 1.40	50 55	
Moisture-, heat-, and oil-resistant thermoplastic	MTW	60°C 140°F	Machine tool wiring in wet locations	Flame-retardant moisture-, heat-, and oil-resistant thermoplastic		(A)	(A)	(A) None (B) Nylon jacket o equivalent
		90°C 194°F	Machine tool wiring in dry locations. FPN: See NFPA 79.		22–12 10 8 6 4–2 1–4/0 213–500 501–1000	0.76 0.76 1.14 1.52 1.52 2.03 2.41 2.79	30 30 45 60 60 80 95 110	
Paper		85°C 185°F	For underground service conductors, or by special permission	Paper				Lead sheath
Perfluoro-alkoxy	PFA	90°C 194°F 200°C 392°F	Dry and damp locations Dry locations — special applications[2]	Perfluoro-alkoxy	14–10 8–2 1–4/0	0.51 0.76 1.14	20 30 45	None
Perfluoro-alkoxy	PFAH	250°C 482°F	Dry locations only. Only for leads within apparatus or within raceways connected to apparatus (nickel or nickel-coated copper only)	Perfluoro-alkoxy	14–10 8–2 1–4/0	0.51 0.76 1.14	20 30 45	None
Thermoset	RHH	90°C 194°F	Dry and damp locations		14–10 8–2 1–4/0 213–500 501–1000 1001–2000	1.14 1.52 2.03 2.41 2.79 3.18	45 60 80 95 110 125	Moisture-resistant, flame-retardant, nonmetallic covering
Moisture-resistant thermoset	RHW[4]	75°C 167°F	Dry and wet locations	Flame-retardant, moisture-resistant thermoset	14–10 8–2 1–4/0	1.14 1.52 2.03	45 60 80	Moisture-resistant, flame-retardant, nonmetallic covering
	RHW-2	90°C 194°F			213–500 501–1000 1001–2000	2.41 2.79 3.18	95 110 125	
Silicone	SA	90°C 194°F	Dry and damp locations	Silicone rubber	14–10 8–2 1–4/0	1.14 1.52 2.03	45 60 80	Glass or other suitabl braid material
		200°C 392°F	For special application[2]		213–500 501–1000 1001–2000	2.41 2.79 3.18	95 110 125	

(Contin

FIGURE 7.3

NEC Table 310.13 (Continued).

Table 310.13(A) *Continued*

Trade Name	Type Letter	Maximum Operating Temperature	Application Provisions	Insulation	Thickness of Insulation AWG or kcmil	mm	mils	Outer Covering[1]
Thermoset	SIS	90°C 194°F	Switchboard wiring only	Flame-retardant thermoset	14–10 8–2 1–4/0	0.76 1.14 2.41	30 45 55	None
Thermoplastic and fibrous outer braid	TBS	90°C 194°F	Switchboard wiring only	Thermoplastic	14–10 8 6–2 1–4/0	0.76 1.14 1.52 2.03	30 45 60 80	Flame-retardant, nonmetallic covering
Extended polytetra-fluoro-ethylene	TFE	250°C 482°F	Dry locations only. Only for leads within apparatus or within raceways connected to apparatus, or as open wiring (nickel or nickel-coated copper only)	Extruded polytetra-fluoroethylene	14–10 8–2 1–4/0	0.51 0.76 1.14	20 30 45	None
Heat-resistant thermoplastic	THHN	90°C 194°F	Dry and damp locations	Flame-retardant, heat-resistant thermoplastic	14–12 10 8–6 4–2 1–4/0 250–500 501–1000	0.38 0.51 0.76 1.02 1.27 1.52 1.78	15 20 30 40 50 60 70	Nylon jacket or equivalent
Moisture- and heat-resistant thermoplastic	THHW	75°C 167°F 90°C 194°F	Wet location Dry location	Flame-retardant, moisture- and heat-resistant thermoplastic	14–10 8 6–2 1–4/0 213–500 501–1000 1001–2000	0.76 1.14 1.52 2.03 2.41 2.79 3.18	30 45 60 80 95 110 125	None
Moisture- and heat-resistant thermoplastic	THW	75°C 167°F 90°C 194°F	Dry and wet locations Special applications within electric discharge lighting equipment. Limited to 1000 open-circuit volts or less. (size 14-8 only as permitted in 410.68)	Flame-retardant, moisture- and heat-resistant thermoplastic	14–10 8 6–2 1–4/0 213–500 501–1000 1001–2000	0.76 1.14 1.52 2.03 2.41 2.79 3.18	30 45 60 80 95 110 125	None
	THW-2	90°C 194°F	Dry and wet locations					
Moisture- and heat-resistant thermoplastic	THWN	75°C 167°F	Dry and wet locations	Flame-retardant, moisture- and heat-resistant thermoplastic	14–12 10 8–6 4–2	0.38 0.51 0.76 1.02	15 20 30 40	Nylon jacket or equivalent
	THWN-2	90°C 194°F			1–4/0 250–500 501–1000	1.27 1.52 1.78	50 60 70	
Moisture-resistant thermoplastic	TW	60°C 140°F	Dry and wet locations	Flame-retardant, moisture-resistant thermoplastic	14–10 8 6–2 1–4/0 213–500 501–1000 1001–2000	0.76 1.14 1.52 2.03 2.41 2.79 3.18	30 45 60 80 95 110 125	None
Underground feeder and branch-circuit cable — single conductor (for Type UF cable employing more than one conductor, see Article 340.)	UF	60°C 140°F 75°C 167°F[6]	See Article 340.	Moisture-resistant Moisture- and heat-resistant	14–10 8–2 1–4/0	1.52 2.03 2.41	60[5] 80[5] 95[5]	Integral with insulation

FIGURE 7.3

NEC Table 310.13 (Continued).

Table 310.13(A)   *Continued*

Trade Name	Type Letter	Maximum Operating Temperature	Application Provisions	Insulation	Thickness of Insulation			Outer Cover
					AWG or kcmil	mm	mils	
Underground service-entrance cable — single conductor (for Type USE cable employing more than one conductor, see Article 338.)	USE	75°C 167°F	See Article 338.	Heat- and moisture-resistant	14–10 8–2 1–4/0 213–500 501–1000 1001–2000	1.14 1.52 2.03 2.41 2.79 3.18	45 60 80 95[7] 110 125	Moisture-resistant nonmetallic cover (See 338.2.)
	USE-2	90°C 194°F	Dry and wet locations					
Thermoset	XHH	90°C 194°F	Dry and damp locations	Flame-retardant thermoset	14–10 8–2 1–4/0 213–500 501–1000 1001–2000	0.76 1.14 1.40 1.65 2.03 2.41	30 45 55 65 80 95	None
Moisture-resistant thermoset	XHHW[4]	90°C 194°F 75°C 167°F	Dry and damp locations Wet locations	Flame-retardant, moisture-resistant thermoset	14–10 8–2 1–4/0 213–500 501–1000 1001–2000	0.76 1.14 1.40 1.65 2.03 2.41	30 45 55 65 80 95	None
Moisture-resistant thermoset	XHHW-2	90°C 194°F	Dry and wet locations	Flame-retardant, moisture-resistant thermoset	14–10 8–2 1–4/0 213–500 501–1000 1001–2000	0.76 1.14 1.40 1.65 2.03 2.41	30 45 55 65 80 95	None
Modified ethylene tetrafluoroethylene	Z	90°C 194°F 150°C 302°F	Dry and damp locations Dry locations — special applications[2]	Modified ethylene tetrafluoroethylene	14–12 10 8–4 3–1 1/0–4/0	0.38 0.51 0.64 0.89 1.14	15 20 25 35 45	None
Modified ethylene tetrafluoroethylene	ZW	75°C 167°F 90°C 194°F 150°C 302°F	Wet locations Dry and damp locations Dry locations — special applications[2]	Modified ethylene tetrafluoroethylene	14–10 8–2	0.76 1.14	30 45	None
	ZW-2	90°C 194°F	Dry and wet locations					

[1] Some insulations do not require an outer covering.

[2] Where design conditions require maximum conductor operating temperatures above 90°C (194°F).

[3] For signaling circuits permitting 300-volt insulation.

[4] Some rubber insulations do not require an outer covering.

[5] Includes integral jacket.

[6] For ampacity limitation, see 340.80.

[7] Insulation thickness shall be permitted to be 2.03 mm (80 mils) for listed Type USE conductors that have been subjected to special investigations. The nonmetallic covering over individual rubber-covered conductors of aluminum-sheathed cable and of lead-sheathed or multiconductor cable shall not be required to be flame retardant. For Type MC cable, see 330.104. For nonmetallic-sheathed cable, see Article 334, Part III. For Type UF cable, see Article 340, Part III.

- **THWN** conductors are PVC-insulated nylon-jacketed conductors that can be used in wet or dry locations.

- **XHHW-2** designates conductor types with heat- and moisture-resistant insulation for use in 90-degree C wet or dry locations.

- **Z** conductors have ETFE insulation and are used in dry locations.

- **ZW** conductors have ETFE insulation and can be used in wet or dry locations.

AMPACITY

NEC Article 310.15 addresses ampacity in great detail. The information covered needs to be cross-referenced to **NEC** Table 310.5, because Article 310.15 doesn't take voltage or voltage drop into account. When you run calculations from the tables, you might come up with more than one ampacity, in which case 310.5(A)(2) instructs you to use the lowest of two or more ampacities. In other words, you have to assume less ampacity than the more favorable calculations demonstrate. Additionally, you have to adjust or "derate" your ampacity calculations based on the number of current-carrying conductors in a wireway, as explained in **NEC** 310.15(B)(2).

When you use Table **NEC** 310.15(B)(2)a. and count the number of current carrying conductors in a raceway, you do not need to count equipment grounding or bonding conductors in your deration of ampacity at all, because they are not current-carrying conductors. The conductor ampacity that you adjusted from the ambient temperature deration calculation has to be multiplied by the percentage of ampacity deration that is required by NEC 310.15(B)(2)a.

Trade Tip

Any conduit that is shorter than 24 inches does not require derating calculations when you are considering the number of conductors installed in a raceway, per *NEC* Article 310.15.B.2. Exception 3.

Number of Current-Carrying Conductors	Values in NEC Tables 310.16 - 310.19 Adjusted for Ambient Temperature, if Necessary
4-6	80 %
7-9	70 %
10-20	50 %
21-30	45 %
31-40	40 %
40 and above	35 %

FIGURE 7.4

Adjustment factors are required for more than three current-carrying conductors in a raceway or rable.

CONDUCTOR LIMITATIONS

NEC 310.15(B)(2)a contains a number of exceptions and requirements. For example, derating is used in the following ways:

- The derating calculation is based on the number of current-carrying conductors installed in the same conduit or in a bundle of cables that is more than 24 inches long.
- Derating also applies to any nonmetallic sheathed cables that are bundled together and any current-carrying conductors installed within a raceway or conduit.

Derating is not used in under the following conditions:

- Short sections of bundled conductors that are less than 24 inches long
- Conductors installed within a nipple that is less than 24 inches long

NEC Article 310.15(B)(2)a and NEC Table 310.16 correction factors do not apply to the residential service-entrance and feeder-conductor ampacity ratings that are provided in NEC Table 310.15(B)6 because this service-entrance and feeder-conductor chart is exempt from both the requirements for ambient temperature and for the number of conductors in a raceway.

CODE UPDATE

A new requirement is a subdivision (c) added to Section
310.15(B)(2) in the 2008 edition of the **NEC**. It includes a new
companion Table 310.15(B)(2)(c). FPN 2 in Section 310.10 was
deleted entirely. It requires ampacity correction due to ambient
temperatures that affect conductors that are exposed to sunlight
on roofs. In electrical installations of conduit, tubing, and cable
that are on rooftops and exposed to sunlight, temperature value
factors in accordance with the new Table 310.15(B)(2)(c) must
now be added. The determining variable is the height that the
wiring method is installed above the roof surface. For example, if a
conduit is installed 12 inches above a rooftop in an ambient
temperature of 122 degrees F, then you have to add 30 degrees to
the anticipated maximum ambient temperature in which the
conduit is installed. In this case, the temperature-correction-factor
adjustments from the applicable table in 310 indicate that you
must assume a temperature of 152 degrees F.

Single-family, two-family, and multifamily dwellings with conductor types
that are listed in **NEC** Table 310.15(B)6 can be used for 120/240 volt,
3-wire, single-phase service-entrance conductors, service laterals, and
feeder conductors that are installed in a raceway or cable with or without
an equipment grounding conductor and that act as the main power feeder
to each dwelling. As a result, the number of conductors is limited to no
more than three. There is also a rule that applies to this chart that requires
any feeder being served by a service-entrance conductor be sized no larger
than the service-entrance conductor.

For example, you may have a situation in which four feeders in a conduit
are served by a service entrance that is sized by Table 310.15(B)6. Their size
also must be determined from this same chart. The result is that the feeder
is not required to be larger than the service-entrance conductors supplying
the power, and thus would be exempt from the deration factors required
by 310.15(B)(2)a or Table 310.16 correction factors.

When counting the total number of current-carrying conductors required by deration of ampacity, you do not need to count any white neutral conductors that are carrying only the unbalanced load of two hot

SERVICE or FEEDER AMPERES	COPPER CONDUCTORS	ALUMINUM OR COPPER-CLAD ALUMINUM CONDUCTORS
100	4	2
110	3	1
125	2	1/0
150	1	2/0
175	1/0	3/0
200	2/0	4/0
225	3/0	250
250	4/0	300
300	250	350
350	350	500
400	400	600

FIGURE 7.5

Approved conductor types are RHH, RHW, RHW-2, THHN, THHW, THW, THW-2, THWN, THWN-2, XHHW, XHHW-2, SE, USE, and USE-2.

conductors on the same two pole circuits and that are installed in the same raceway. An example of this scenario would be a multiwire circuit comprised of two 120-volt branch circuits using the same common white wire to a 240-volt receptacle box. In this case, there would be two hot conductors, one insulated white neutral conductor, and one equip-ment-grounding conductor. Another example would be the neutral conductor in an electric range that uses both 240 volts and 120 volts in the same multiwire branch circuit.

You have to count the white grounded leg serving a single ungrounded hot conductor in a 120-volt branch circuit in conduit-fill calculations. This type of white conductor, found for example as the white wire in a 120-volt receptacle, is not a true neutral conductor. A white wire that is a neutral must be a conductor that serves two ungrounded conductors that are

Trade Tip

A 120-volt branch circuit serving receptacles is not included in this rule for neutrals, because it does not have a neutral conductor in its branch-circuit design.

Trade Tip

Conductor ampacity is the ampacity times the correction factor for ambient temperature times the correction factor for the number of conductors in a conduit over three conductors. This equates to the adjusted conductor ampacity, which is recognized as the ampacity of that conductor by the **NEC**.

served from different phases. Those two ungrounded conductors that are served by a neutral conductor must have a voltage of at least 200 volts or more on a multiwire feeder or branch circuit that is made up of two hot conductors from different phases. These two ungrounded conductors must measure voltage between the two ungrounded conductors and use a double or triple pole breaker or set of fuses. No single 120-volt circuit is served by a neutral conductor. **NEC** 310.15(B)4 b requires that the neutral of a 4-wire 3-phase wye-connected conductor be counted as a current-carrying conductor.

SINGLE-WIREWAY INSTALLATION

NEC Table 310.16 applies to situations where you have three or less current-carrying conductors in a single wireway. This is not very unusual for services and feeders but is not typical of branch circuits. First, determine whether you will be using copper or aluminum, then select the corresponding conductor type from the column that shows the cable designations, based on the temperature rating of the conductors, which were originally determined using **NEC** Table 310.13.

Did You Know?

NEC 210.14.C.1 and 2 require utilizing a lower ampacity rating than found in **NEC** Table 310.16 because the terminal ends and wire nuts that are sold on the market in the electrical industry are only rated at 60 degrees C if they are for #1 AWG and only 75 degrees C if they are for larger than #1 AWG conductors, as per **NEC** Article 110.14.C.1 and 2.

TEMPERATURE RATING OF CONDUCTOR

AWG Size	COPPER 60°C (140°F) Types TW, UF	75°C (167°F) Types RHW, THW, THHW, THWN, XHHW, USE, ZW	90°C (194°F) Types TBS, SA, SIS, FEP, FEPB, MI, RHH, RHW-2, THHN, THHW, THW-2, THWN-2, USE-2, XHH, XHHW, XHHW-2, ZW-2	ALUMINUM OR COPPER-CLAD ALUMINUM 60°C (140°F) Types TW, UF	75°C (167°F) Types RHW, THW, THWN, XHHW, USE	90°C (194°F) Types TBS, SA, SIS, THHN, THHW, THW-2, THWN-2, RHH, RHW-2, USE-2, XHH, XHHW, XHHW-2, ZW-2	AWG Size
18	–	–	14	–	–	–	–
16	–	–	18	–	–	–	–
14*	20	20	25	–	–	–	–
12*	25	25	30	20	20	25	12*
10*	30	35	40	25	30	35	10*
8	40	50	55	30	40	45	8
6	55	65	75	40	50	60	6
4	70	85	95	55	65	75	4
3	85	100	110	65	75	85	3
2	95	115	130	75	90	100	2
1	110	130	150	85	100	115	1
1/0	125	150	170	100	120	135	1/0
2/0	145	175	195	115	135	150	2/0
3/0	165	200	225	130	155	175	3/0
4/0	195	230	260	150	180	205	4/0
250	215	255	290	170	205	230	250
300	240	285	320	190	230	255	300
350	260	310	350	210	250	280	350
400	280	335	380	225	270	305	400
500	320	380	430	260	310	350	500
600	355	420	475	285	340	385	600
700	385	460	520	310	375	420	700
750	400	475	535	320	385	435	750
800	410	490	555	330	395	450	800
900	435	520	585	355	425	480	900
1000	455	545	615	375	445	500	1000
1250	495	590	665	405	485	545	1250
1500	520	625	705	435	520	585	1500
1750	545	650	735	455	545	615	1750
2000	560	665	750	470	560	630	2000

CORRECTION FACTORS

For ambient temperatures other tha[n] 30°C (86°F), multiply the allowable ampacities shown above by the appropriate factor shown below.

Ambient Temp. (°C)							Ambient Temp. (°F)
21-25	1.08	1.05	1.04	1.08	1.05	1.04	70-77
26-30	1.00	1.00	1.00	1.00	1.00	1.00	78-86
31-35	0.91	0.94	0.96	0.91	0.94	0.96	87-95
36-40	0.82	0.88	0.91	0.82	0.88	0.91	96-104
41-45	0.71	0.82	0.87	0.71	0.82	0.87	105-113
46-50	0.58	0.75	0.82	0.58	0.75	0.82	114-122
51-55	0.41	0.67	0.76	0.41	0.67	0.76	123-131
56-60	–	0.58	0.71	–	0.58	0.71	132-140
61-70	–	0.33	0.58	–	0.33	0.58	141-158
71-80	–	–	0.41	–	–	0.41	159-176

FIGURE 7.6

Allowable ampacities of 0- to 2000-volt insulated conductors with no more than three current-carrying conductors in a raceway or cable.

If you are required to perform an ampacity deration calculation for either ambient temperature or more than three current-carrying conductors in a raceway, start your calculation using the ampacity rating in the column for 90 degrees C in Figure 7.6 if the conductor's insulation was originally rated for 90 degrees. You can start your derating calculations by using the ampacity rating found in the column for 75 degrees C in Figure 7.6 if the conductor's insulation was originally rated for 75 degrees. Next, you have to finish your derating calculation for the ampacity rating based on the requirements of **NEC** 110.14.C.

Once you have completed your ampacity deration calculation, compare the derated ampacity number you came up with to the original temperature-rated ampacity for either the 60-degree ampacity rating (if the conductor is smaller than 1 AWG) or the 75-degree ampacity rating (if the conductor is 1 larger than 1 AWG). Next, you are required to use the worst-case scenario, which means the lowest ampacity rating of either your derated ampacity or the ampacity rating using **NEC** 110.14.C.1 and C2.

APPLYING DERATING TO UP TO THREE CONDUCTORS

So far we have looked at a lot of tables and derating basics. Let's practice what we've learned with an example of conductors that will run through the attic of a house in Texas. Here is the essential installation information you need to begin:

- We are running a #8 AWG TW copper conductor in an attic that has an approximate average ambient temperature of 130 degrees F. There will be a total of 26 current-carrying conductors in the conduit.
- STEP 1: Begin by looking at **NEC** Table 310.16 in the first column for 60° C/140° F). You will see that #8 copper TW is rated at 40 amperes.

Trade Tip

Anytime you have an ampacity that is 0.5 or less, you need to round down to the next whole number. In the case of our example, round 7.38 down to 7.

- STEP 2: In order to determine what correction factor to use, you will need to convert 130° F to C. The formula for this is:

$$C = (F - 32) \div 9 \times 5$$

  Using this formula, your conversion would work like this:

  $$130 - 32 = 98 \quad 98 \div 9 = 10.888 \quad 10.9 \times 5 = 54.5$$

- STEP 3: Locate 55°C in the ambient-temperature column of the correction-factors section of Table 310.16. You will see that the corresponding correction factor is 0.41.

- STEP 4: Multiply 40 amperes from the table by the 41 percent correction factor above, which will give you a total of 16.4. This means that you have 16.4 amperes of ampacity after the temperature-deration calculation.

- STEP 5: Return to Table 310.15(B)(2) and locate the number of current-carrying conductors in your raceway, which we listed in the essential installation information as 26 conductors, and find the adjustment value for ambient temperature that correlates to the quantity of conductors. For between 21 and 30 conductors, you will use the value of 45 percent.

- STEP 6: Next, multiply the calculated ampere result of 16.4 times the 45 percent value: $16.4 \times 0.45 = 7.38$. This means that 7.38 is the total reduction of ampacity after the derations for ambient temperature and quantity of conductors are factored in. Round down to 7. Now you have determined that 7 amperes is the true adjusted total ampacity rating of the #8 AWG TS copper conductor. But wait, there is more.

- STEP 7: The **NEC** requires you to have a minimum ampacity of 15 amps for any conductor that is used in a wiring system of 120 volts or more (Table 210.3). Furthermore, the conductor has to have an ampacity that is no less than the maximum load to be served, based on the requirements of **NEC** 210.19(A)1.

- CONCLUSION: You just ran all these calculations only to find that the #8 AWG TW conductor you planned to use in this installation is not permitted for this use. Not only is it not rated at the minimum required 15 amperes, but because it is smaller than 15 amperes it is too small to carry the 120-volt service load.

TEMPERATURE RATING OF CONDUCTOR

AWG Size	COPPER			ALUMINUM OR COPPER-CLAD ALUMINUM			AWG Size
	60°C (140°F) Types TW, UF	75°C (167°F) Types RHW, THHW, THW, THWN, XHHW, USE, ZW	90°C (194°F) Types TBS, SA, SIS, FEP, FEPB, MI, RHH, RHW-2, THHN, THHW, THW-2, THWN-2, USE-2, XHH, XHHW, XHHW-2, ZW-2	60°C (140°F) Types TW, UF	75°C (167°F) Types RHW, THHW, THW, THWN, XHHW	90°C (194°F) Types TBS, SA, SIS, THHN, THHW, THW-2, THWN-2, RHH, RHW-2, USE-2, XHH, XHHW, XHHW-2, ZW-2	
18	–	–	18	–	–	–	–
16	–	–	24	–	–	–	–
14*	25	30	35	–	–	–	–
12*	30	35	40	25	30	35	12*
10*	40	50	55	35	40	40	10*
8	60	70	80	45	55	60	8
6	80	95	105	60	75	80	6
4	105	125	140	80	100	110	4
3	120	145	165	95	115	130	3
2	140	170	190	110	135	150	2
1	165	195	220	130	155	175	1
1/0	195	230	260	150	180	205	1/0
2/0	225	265	300	175	210	235	2/0
3/0	260	310	350	200	240	275	3/0
4/0	300	360	405	235	280	315	4/0
250	340	405	455	265	315	355	250
300	375	445	505	290	350	395	300
350	420	505	570	330	395	445	350
400	455	545	615	355	425	480	400
500	515	620	700	405	485	545	500
600	575	690	780	455	540	615	600
700	630	755	855	500	595	675	700
750	655	785	885	515	620	700	750
800	680	815	920	535	645	725	800
900	730	870	985	580	700	785	900
1000	780	935	1055	625	750	845	1000
1250	890	1065	1200	710	855	960	1250
1500	980	1175	1325	795	950	1075	1500
1750	1070	1280	1445	875	1050	1185	1750
2000	1155	1385	1560	960	1150	1335	2000

CORRECTION FACTORS

For ambient temperatures other than 30°C (86°F), multiply the allowable ampacities shown above by the appropriate factor shown below.

Ambient Temp. (°C)							Ambient Temp. (°F)
21-25	1.08	1.05	1.04	1.08	1.05	1.04	70-77
26-30	1.00	1.00	1.00	1.00	1.00	1.00	78-86
31-35	0.91	0.94	0.96	0.91	0.94	0.96	87-95
36-40	0.82	0.88	0.91	0.82	0.88	0.91	96-104
41-45	0.71	0.82	0.87	0.71	0.82	0.87	105-113
46-50	0.58	0.75	0.82	0.58	0.75	0.82	114-122
51-55	0.41	0.67	0.76	0.41	0.67	0.76	123-131
56-60	–	0.58	0.71	–	0.58	0.71	132-140
61-70	–	0.33	0.58	–	0.33	0.58	141-158
71-80	–	–	0.41	–	–	0.41	159-176

FIGURE 7.7

Use this chart to determine correction factors for branch circuits.

Did You Know?

Even if you are not required to calculate for derations, are using a higher temperature-rated conductor such as THHN, which is rated 90 degrees C in **NEC** Table 310.16, and are working with a conductor that is smaller than #1 AWG, you still must use the 60-degree C column as the ampacity rating. This is in order to comply with 110.14(C)1.a.1, which requires that the temperature rating for a conductor cannot exceed the lowest temperature rating of any termination, conductor, or device connected to it.

SINGLE-INSULATED CONDUCTORS

The next important section of Article 310 is **NEC** Table 310.17. This table is used for installations that use single-insulated conductors. Don't let the name fool you. This doesn't mean that you use only one conductor—it refers to conductors that aren't insulated twice, such as a conductor that is in an insulated sheath with other conductors in free air. It is a slightly convoluted way of describing conductors typically used for branch circuits. Although the table is designed in the same format as Table 310.16, to use this table you start by selecting the conductor material, either copper or aluminum, and then use the column that shows the cable type, which is listed by the insulating-material letter designation, such as THHN, which has PVC-insulated nylon jacketing.

AT A GLANCE

While the tables in **NEC** 310.16 through 310.18 may all look the same at first glance, the differences between them are clear:

- **NEC** Table 310.16 and **NEC** Table 310.17 are the basic ampacity tables.
- **NEC** Table 310.16 is typically for service entrances.
- **NEC** Table 310.17 is typically for branch circuits.

- NEC Table 310.18 is the equivalent of **NEC** Table 310.16, but for installations at higher temperatures.

- NEC Table 310.19 corresponds to **NEC** Table 310.17 except that it is used for installation at higher temperatures.

The balance of **NEC** Article 310 is dedicated to the cable-installation diagrams provided in **NEC** Figure 310.60, which are used for reference for the remaining conductor-application tables that provide the standards for various conductor sizes and types.

8

Boxes, Cables, Wireways, and Raceways

WHAT YOU NEED TO KNOW

This chapter's title may look daunting, but these sections of the code are grouped together because they are so strongly associated with each other. Beginning at the panelboard, you need to get a cable to an outlet, which will have to run through wireways, raceways, or conduit along its journey. As a professional electrician, you are expected to know how to install, route, splice, protect, and secure conductors and raceways. Additionally, the various conditions of use for each type and location of material to be installed are critical. If there is one section of the code that you absolutely need to know in order to perform electrical work, it is Chapter 3 (**NEC** 312-392). The articles in this section cover everything from junction boxes and cable assemblies, to metal and PVC conduits, busways, cablebuses, and raceways. There are over 60 tables packed full of temperature ratings, bending-space and box-size requirements, ampacity and conduit sizing, and allowable fill areas. With so much information to cover, we are going to condense as many facts and requirements into this chapter as possible.

TERMS TO KNOW

Box: A metallic or nonmetallic receptacle designed for connections to a wiring system.

Busway: A metal enclosure that is grounded and has factory-mounted, bare, or insulated conductors that are usually copper or aluminum rods, bars, or tubes.

Cable Bus: A completely enclosed assembly of insulated conductors, fittings, and conductor terminations run in a ventilated protective metal housing.

Raceway: An enclosed conduit that forms a physical pathway for electrical wiring.

Wireway: A complete raceway system.

CABINETS AND CUTOUT BOXES

Cabinets

An electrical cabinet has to be installed in drywall or plaster so that the front edge doesn't set back further than 6 inches from the finished surface of the wall. If the cabinet is installed in wood or a combustible material, then it has to be flush with the finished surface (**NEC** 312.3).

Anywhere that conductors enter the cabinet, the openings have to be adequately closed (**NEC** 312.5A). Conductors in an enclosure for switches or overcurrent devices cannot fill the wiring space by more than 40 percent of the cross-section area inside the enclosure, and they can't fill the cross-sectional area of the space by more than 75 percent.

Minimum wire-bending requirements at terminals are listed in **NEC** Tables 312.6(A) and 312.6(B).

Trade Tip

There must be at least 1 inch of airspace between a cabinet door and any live metal parts, including fuses (**NEC** 312.11(A)2).

Outlets and Junction Boxes

Round boxes cannot be used if conduits or conductors connected to the sides of the boxes would require locknuts or bushings (**NEC** 314.2).

NEC Table 314.16(A) provides the box size and maximum number of conductors permitted in metal boxes. **NEC** Article 314.4 requires all metal boxes to be grounded (bonded), and **NEC** Article 314.5 forbids you from putting splices in short-radius conduit bodies.

Box-fill calculations are required based on the number of conductors and the sizes of any cable clamps located inside a box. Every conductor that originates outside the box and terminates inside the box is counted as one conductor (**NEC** 314.16(B)1 and 2). A conductor that originates outside the box and is spliced inside the box counts as a total of two conductors. Conductors that originate outside o the box and are terminated inside the box with a wire connector count as two conductors, even though they are connected.

FIGURE 8.1

Many metal boxes have "knockouts" so you can install up to a specific quantity of conductors without drilling.

Look up the box size and maximum number of conductors, by conductor size, in **NEC** Table 314.16(A), to determine the total quantity of conductors permitted in a box.

Outlet boxes that are used to support light fixtures can only be installed for luminaires (light fixtures) that weigh 50 pounds or less (**NEC** 314.27 B). Outlet boxes that will be the sole support for ceiling paddle-fan installations can only be installed for fans that weigh 70 pounds or less (**NEC** 314.27 D). A new last sentence has been added to the end of **NEC** 314.27 D in the 2008 edition; it specifies the type of box that must be used if two or more switched ungrounded conductors or switch legs are roughed into a paddle-fan outlet box. You are now required to install a box that is specifically listed for the sole support of a ceiling-suspended paddle fan.

FIGURE 8.2

*Based on the requirements of **NEC** 314.16, this box contains five conductors.*

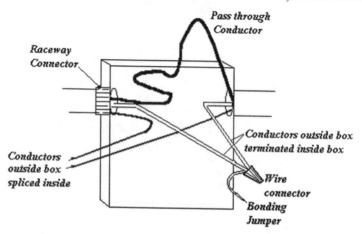

CODE UPDATE

A new section has been added as **NEC** 314.27 in the 2008 edition for boxes that are used to support utilization equipment other than ceiling-suspended (paddle) fans. These boxes must meet the requirements of 314.27 (A) and (B) for the support of a light fixture of the same size and weight. An exception has also been included for utilization equipment that weighs 6 pounds or less and that can be supported on other boxes and secured to these box with at least two No. 6 or larger screws. This change provides clarification for the installation of boxes for equipment such as smoke detectors that need to be mounted on an outlet box.

Boxes and conduit can be used as junction or pull boxes (**NEC** 314.28), as long as conductors in the raceway are #4 AWG or larger and meet the following requirements:

- For straight pulls the length of the box must be at least eight times the trade size of the largest raceway.
- For angle pulls, if splices, angles, or "U" pulls are made, the distance between the location where each raceway enters the box and the opposite side of the box cannot be less than six times the trade size of the largest raceway in a row or of the largest raceway.

Metal boxes must provide a way to connect an equipment grounding conductor to the box. A tapped hole is considered an acceptable means for this connection (**NEC** 314.40 D).

Trade Tip

Pull boxes and junction boxes that exceed 6 feet in any dimension have to contain conductors that are cabled or racked, per **NEC** 314.28(B).

CABLE TYPES

NEC Articles 320 through 340 define the standards and requirements for various types of cables.

Armored Cable

Armored cable (**Type AC**) is an assembly of insulated conductors, in sizes 14 AWG through 1 AWG, that are individually wrapped within waxed paper (Article 320). The conductors must be enclosed in a flexible spiral metal sheath of either steel or aluminum that interlocks at the edges. Although armored cable has an outward appearance similar to flexible metal conduit, it is not the same.

Metal-Clad Cable

Metal-clad cable (Type MC) encloses one or more insulated conductors in a metal sheath that is made of either corrugated or smooth copper or aluminum tubing (Article 330). MC can also have a spiral interlocked steel or aluminum sheath. Due to this construction, MC cable is very versatile and can be used for almost any application and in a number of specified locations.

Nonmetallic-Sheathed Cable

Nonmetallic-sheathed cable (Type NM) encloses two or three insulated conductors, sizes 14 AWG through 2 AWG, with a nonmetallic outer cover, and it contains a separate equipment grounding conductor (Article 334). Nonmetallic-sheathed cable is commonly used for residential and commercial branch circuits. You cannot use NM cable for service wiring or for outside applications unless otherwise permitted.

Fast Fact

There are no limitations to the number of bends between terminations when you use armored cable.

FIGURE 8.3

Metal-clad cable has a metal sheath and outer covering.

Power and Control Tray Cable

Tray cable (Type TC) is a factory-assembled cable that consists of two or more insulated conductors enclosed in a nonmetallic sheath (Article 336). It is used in cable trays, raceways, or where it is supported by a messenger wire, but it cannot be used as open cable on brackets or cleats.

FIGURE 8.4

Nonmetallic-sheathed cable assemblies include a separate grounding conductor.

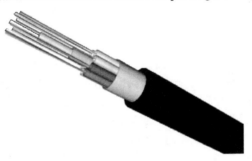

Trade Tip

You are permitted to use SE and USE for feeders and branch circuits.

Service-Entrance Cable

Service-entrance cable (Types SE and USE) can have either a single conductor or multiconductor cable-covered assembly and is primarily used for electrical services that are 600 volts or less (Article 338). Anytime you use SE or USE cable, it must be installed per the requirements of **NEC** 230.

Underground Feeder Cable

Underground feeder cable (Type UF) is a moisture- and corrosion-resistant cable that can be used in direct buried installations and comes sizes that range from 14 AWG to 4/0 AWG (Article 340). It has a molded plastic multiconductor covering and insulated conductors. UF cannot be used for service wiring.

CONDUIT

Conduit that encloses #6 AWG or smaller-sized conductors needs to have a cross-sectional area that is not less than two times the cross-sectional area of the largest conduit that it is attached to (**NEC** 314.16 C). The maximum number of conductors that can be run in conduit is listed in **NEC** Annex C by conduit type. Conduit runs are also called raceways. Descriptions of the various types of conduit follow.

Intermediate Metal Conduit

Intermediate metal conduit (Type IMC) is a round metal raceway with an outside diameter that matches rigid metal conduit (RMC) but that has a wall thickness that is less than RMC (Article 342). This means that it has a greater interior cross-sectional area. IMC is made of a different steel alloy than RMC, which makes it more rigid, even though the wall thickness is smaller.

Trade Tip

Intermediate metal conduit is lighter in weight and less expensive than RMC and can be used in the same locations, which makes it a very desirable raceway choice.

Rigid Metal Conduit

Rigid metal conduit (Type RMC) is a round metal raceway with a wall thickness that is greater than MC, which means that it has a smaller interior cross-sectional area (Article 344). RMC can be used in any location.

Flexible Metal Conduit

Flexible metal conduit (Type FMC) is a round raceway that is made of formed, interlocked metal strips of either steel or aluminum that are wound in the shape of a helix (Article 348). The common term for FMC is "flex conduit," and it is usually installed in the last 6 feet or less of raceway between a more rigid raceway and equipment that moves or vibrates, such as a pump motor. **NEC** Table 348.22 lists the maximum number of insulated AWG conductors that are permitted to be used with FMC.

FIGURE 8.5

Rigid metal conduit is generally threaded to attach lengths of conduit or connectors.

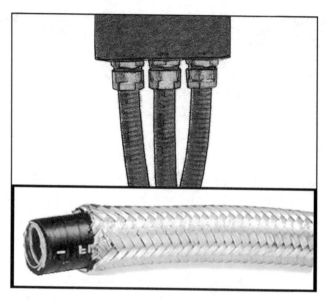

FIGURE 8.6

Flexible metal conduit provides a bendable connection between a rigid raceway and equipment or machinery that generates vibration.

Liquidtight Flexible Metal Conduit

Lightweight flexible metal conduit (Type LFMC) is a listed circular raceway that has an outer nonmetallic, thermoplastic, sunlight-resistant jacket that encases a flexible inner metal core and is liquid-tight (Article 350). It is installed using couplings and connectors and is approved for the installation of electric conductors. Common names for LFMC are the brand name Sealtight® or the generic term "liquid-tight." The smallest allowable trade size is 1/2 inch, and the largest allowable size is 4 inches.

CODE UPDATE

The title of Article 352 has been changed in the 2008 edition of the **NEC** from "Nonmetallic Conduit: Type RNC" to "Polyvinyl Chloride: Type PVC."

FIGURE 8.7

An example of 1/2-inch liquid-tight flexible metal conduit with a flexible metal core.

Polyvinyl Chloride

Polyvinyl chloride conduit (Type PVC) is a round, nonmetallic raceway that utilizes glue-on couplings and connectors and can used for direct burial and underground use encased in concrete as well as in wet locations (Article 352). Schedule 80 PVC can be used in areas where it is subject to physical damage.

High-Density Polyethylene Conduit

High-density polyethylene conduit (Type HDPE) is lightweight and durable and resists decomposition, oxidation, and hostile elements that cause damage to other materials (Article 353). It is used for communications, data, cable television, and general-purpose raceways.

FIGURE 8.8

PVC conduit running into a manhole for direct burial.

CODE UPDATE

Prior to the 2005 publication of the **NEC**, Article 352 included RNC, PVC, RTRC, and HDPE conduits. In the 2005 **NEC**, HDPE was removed from **NEC** 352 and added as Article 353. In the 2008 edition, requirements for PVC and RTRC conduit have been separated, and a new definition of PVC)has been provided in Article 352. A new, separate Article 355 has been added to cover RTRC.

Nonmetallic Underground Conduit with Conductors

Nonmetallic underground conduit with conductors (Type NUCC) is a factory assembly of conductors or cables inside a nonmetallic, smooth-wall conduit with a circular cross section (Article 354). Nonmetallic conduit is manufactured from a material that is resistant to moisture and corrosive agents. It is also capable of being supplied on reels without damage or distortion and is of sufficient strength to withstand abuse, such as impact or crushing in handling or during installation without damage to conduit or conductors. You cannot use Type NUCC in exposed locations or inside buildings.

Reinforced Thermosetting Resin Conduit

Prior to the 2005 publication of the **NEC**, Article 352 included Rigid Non-metallic Conduit (RNC), PVC, RTRC, and HDPE conduits. In the 2005 **NEC**, High Density Polyethylene Conduit (HDPE) was removed from **NEC** 352 and added as Article 353. In the 2008 edition of the **NEC**, requirements for PVC and RTRC conduit have been separated and new definition of Rigid Polyvinyl Chloride Conduit (PVC) has been provided in Article 352. A new, separate Article 355 has been added to cover Reinforced Thermosetting Resin Conduit (RTRC).

Liquid-Tight Flexible Nonmetallic Conduit

Liquid-tight flexible nonmetallic conduit (Type LFNC) is a circular raceway with an outer liquid-tight, nonmetallic, sunlight-resistant jacket and an inner flexible core (Article 356). There are three types of LFNC:

- Type LFNC-A, which has a smooth, seamless inner core and cover that are bonded together with one or more reinforcing layers between the core and the covers

- Type LFNC-B has a smooth inner surface and reinforced walls
- Type LFNC-C has a corrugated external and internal surface but no integral reinforcement in the conduit wall

Electrical Metallic Tubing

Electrical metallic tubing (Type EMT) is a metallic tubing that is not threaded and uses connectors and couplings that are threadless. It can be used in exposed and concealed installations as well as wet locations, but not where it would be exposed to severe physical damage (Article 358).

Electrical Nonmetallic Tubing

Electrical nonmetallic tubing (Type ENT) is a pliable, corrugated raceway that is made of polyvinyl chloride and is commonly referred to as "Smurf" or "Smurf pipe" because, when it originally came out, it was available only in blue (Article 362). It can be used in the walls, floors, and ceilings of buildings that do not have more than three floors or in buildings with more than three floors if a thermal barrier is installed.

BUSWAYS AND CABLEBUSES

Busway installations are described in **NEC** Article 368. An electrical busway system is typically made up of a number of pieces of track that are connected end to end with one or more electrically isolated, conductive

FIGURE 8.9

EMT comes in sizes up to 6 inches, which allows it to carry more conductors in one raceway than many other materials.

> **CODE UPDATE**
>
> A new exception, Number 3, was added to **NEC** 362.30(A) in the 2008 edition. This exception clarifies that, if securing ENT in prefinished wall panels is not practical, you can use unbroken lengths of ENT that do not have couplings.

busbars fastened to the housing. This allows the system to conduct electricity end to end through the busbars, and the busbars are adapted to provide electrical power tap-off at any point along their length. Feeder and branch-circuit busways have to be protected against overcurrent, per **NEC** 368.12. Additionally, **NEC** 368.60 requires busways to be grounded. If a neutral bus is installed, it has to be sized based on the neutral conductor load current, including any harmonic currents, and must have sufficient momentary and short-circuit ratings, per **NEC** 368.258.

NEC 370 cites that a cablebus can be used for branch circuits, feeders, and services, and the conductor ampacity has to be sized based on **NEC** Tables 310.17 and 310.19 up to 600 volts and on Tables 310.69 and 310.70 for over 600 volts.

FIGURE 8.10

A metal busway.

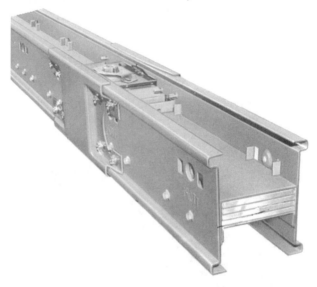

FIGURE 8.11

A cablebus with insulated conductors run through the metal housing.

WIREWAYS

A wireway is a complete raceway system that can be installed in any length. It is a sheet-metal trough, with a hinged or removable cover, that is used to protect electric wires and cables that are installed in it after the wireway has been put in place. **NEC 376.12** prohibits you from installing wireways in severely corrosive environments or where they could be subjected to extreme physical damage. However, they can be used for exposed work and as extensions that pass transversely and unbroken through walls, as long as access to the conductors can be made on both sides of the wall. Based on **NEC 376.22**, the sum of the cross-sectional areas of the total number of conductors permitted in a wireway cannot exceed 20 percent of the interior cross-sectional area of the wireway. If more than 30 conductors, including any neutral conductors, will be installed, then you have to apply derating factors from **NEC 310.15(B)(2)a**.

Nonmetallic wireways are flame-retardant as well as nonmetallic. Based on **NEC 378.12**, they cannot be used in areas that are exposed to sunlight unless they are specifically listed and marked as suitable or are used with conductors that have a higher insulation rating than that of the wireway. As with metal wireways, the sum of the cross-sectional areas of the total number of conductors permitted in a wireway cannot exceed 20 percent of the interior cross-sectional area of the wireway. The derating factors from **NEC 310.15(B)(2)a** have to be applied to the quantity up to and including this 20 percent fill.

Conductors for signaling circuits or controller conductors between a motor and a starter used only to start the motor are not counted as conductors.

RACEWAYS

Raceways protect wires and cables from heat, humidity, corrosion, water intrusion, and physical damage. A strut-type channel raceway is made of metal and is attached to the surface of a structure or suspended from it. Per **NEC** 384, channel-strut raceways can be used in the following installations:

- Exposed locations
- Wet locations
- Areas that are subjected to corrosive vapors, as long as the raceway has a protective finish
- Installations using 600 volts or less
- Power poles
- Class 1, Division 3 hazardous locations (see **NEC** 500.5(B)2)
- Equipment grounding conductors in accordance with **NEC** 250.118(4)

The calculation method for determining the number of conductors allowed in raceways is provided in **NEC** 384.22. The number of wires (n) is equal to the channel area in square inches (ca) divided by the wire area (wa). Provided in Table 384.22 is the inside diameter area of strut-type raceways, based on the size of the raceway.

Surface Metal Raceways

Surface metal raceways are a type of metallic raceway that is mounted to the surface of a structure and can be used in dry locations, Class I, Division 2 installations, and under raised floors (**NEC** 386). They cannot be used

Trade Tip

Ferrous-channel raceways, which are made of iron and protected only by enamel, can only be used outdoors.

Did You Know?

Strut-type channel raceways cannot be installed in concealed locations.

---

where they might be exposed to severe damage or corrosive vapors or installed so that they are concealed. You can generally only use them for installations of 300 volts or less, unless the raceway metal is at least 0.040 inch thick. If the surface metal raceway is used to provide a transition from other kinds of wiring methods, then it must have a means of connecting an equipment grounding conductor.

Surface Nonmetallic Raceways

Surface nonmetallic raceways are mounted to structures in much the same way as metallic raceways and include any associated boxes or connectors needed for the electrical conductors that run in the raceway (**NEC** 388). Primarily used in dry locations, nonmetallic raceways cannot be concealed or used to carry 300 volts or more between conductors unless they are specifically rated for higher voltages.

Underfloor Raceways

Underfloor raceways are used underneath a flooring material, including concrete (**NEC** 390). There are a number of requirements for the amount of cover that must exist over underfloor raceways. For example, half-round and flat-topped raceways that are 4 inches wide or smaller must not have less than 3/4 inch of concrete or wood covering. If the raceway is between 4 inches and 8 inches, then it must have at least 1 inch of space between it and the next raceway, as well as at least 1 inch between the top of the raceway and a concrete floor surface. If the raceways are spaced less than 1 inch apart, then they have to be covered by at least 1 1/2 inches of concrete. The maximum number of conductors permitted in this type of raceway is calculated by combining the cross-sectional area of all the conductors and multiplying that total by 40 percent of the interior cross-sectional area of the raceway.

CABLE TRAYS AND LADDERS

Cable tray systems are used to secure or support cables and raceways (**NEC** 392). The types of cable that are permitted to run in cable-tray systems are listed in **NEC** Table 392.3(A). Cable trays cannot be used in ducts, plenums, or other air-handling spaces unless they are used to support wiring methods, as outlined in **NEC** 300.22. Smooth edges and corrosion protection are required for cable trays, and metallic cable trays have to be grounded. **NEC** Table 392.7(B) lists the metal area required for cable trays when they are used as equipment grounding conductors. Article 392 provides voltage limitations, based on cable types and whether the cable tray or ladder is solid or vented. It also defines whether cables are required to be run in a single layer or can be bundled together.

Once you are familiar with these various means of running cables to their intended locations, you will be ready to install specific equipment and fixtures with confidence.

9

Switches and Lighting Requirements

WHAT YOU NEED TO KNOW

This section of the **NEC** contains a large quantity of lighting information, divided into ten sections. If you picture a lighting installation in your mind, it would follow the sections closely. First you would determine which kind of lights you were going to install and how you would attach them to the ceiling or walls where they are needed. Then you would have to attach them to boxes with adequate supports to hold the fixtures in place. You would need to make sure that you provided adequate grounding or bonding when you run wires to the boxes, and then you would determine the conductors to use. Proper switching, as outlined in **NEC 404**, would be necessary for the operation of the lights, and following any manufacturer's installation requirements would be mandatory. Since there are so many types of lights and locations that require lighting, you would need a thorough understanding of the standards that apply to various kinds of lights, such as track or recessed fixtures. This is the same progression that is provided in **NEC 410**. One important change in the 2008 edition of the **NEC** is that Article 410 has been completely renumbered to allow space for future revisions. It also consolidates definitions into a single section and re-moves all references in parentheses to "light fixtures."

TERMS TO KNOW

Cove Lighting: A light effect, not a specific type of fixture, in which the fix-ture is installed in a manner so that it is directed upward. The opposite ef-fect, known as cornice lighting, directs the light downward.

Electric-Discharge Light: Also known as gaseous discharge, a type of lamp that produces light by establishing a permanent electric arc in a gas. This is the process used to produce light in fluorescent and high-intensity dis-charge lamps.

Lampholder: A device that mechanically supports a bulb (lamp) so that it makes electrical contact with the lamp.

Luminaire: Also known as a light fixture, any of a variety of types of elec-trical devices used to create artificial light or illumination.

Recessed Fixture: A light fixture installed in a protective housing that is concealed behind a ceiling or wall, leaving only the fixture itself exposed.

FIGURE 9.1

Cove fixtures must have adequate space and be properly installed and maintained.

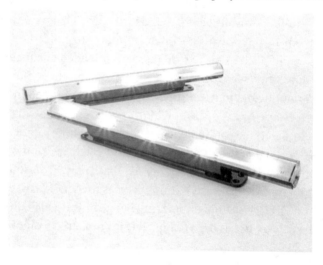

Recessed fixtures that are ceiling-mounted are often called "downlights."

Surface-Mount Fixture: A light fixture with a finished housing that is exposed instead of being installed flush with surface.

Switch: A device rated in amperes, used in general distribution and branch circuits, and capable of interrupting its rated current at its rated voltage.

Track-Light Fixture: A fixture, called a track head, that can be positioned anywhere along a channel track that provides power.

CONNECTING SWITCHES

Three-way and four-way switches must be wired so that all the switching is done in the ungrounded circuit conductor, based on **NEC** 404.2. Switches or circuit breakers cannot be used as a means to disconnect the grounded conductor of a circuit, with the exception of installations in which they disconnect all circuit conductors simultaneously. If metal raceways or metal-armored cables are used in an installation, the wiring between the switches and outlets needs to be done in accordance with **NEC** 300.20(A).

A knife switch has a metal lever and an insulated free end that makes contact with a metal slot. The electrical connections are exposed, which prohibits these kinds of switches from being used in household wiring. Single-throw knife switches have to be installed in a manner that doesn't allow gravity alone to close them. **NEC** 404.6 allows single-throw knife switches with an integral mechanical means that ensures that the blades remain in the open position when the switch is set. Single-throw knife switches and switches with butt contacts have to be installed so that the switch blades are de-energized when the switch is in the open position.

Single-throw knife switches, bolted pressure-contact switches, molded case switches, switches with butt contacts, and circuit breakers are approved switching methods as long as the terminals supplying the switch loads are deenergized when the switch is in the open position.

Fast Fact

Switch loops do not require a grounded conductor.

FIGURE 9.2

Single-throw knife switches must be installed in an inverted position.

Double-throw knife switches can be mounted in a vertical or horizontal throw position, but if the throw is vertical, the switch must have a means to hold the blades in the open position.

A general-use snap switch can be installed in device boxes and used in conjunction with wiring systems. Faceplates for snap switches that are mounted in boxes need to be installed so that they are flush-mounted and completely cover the box opening, per **NEC** 404.9. Snap switches, including dimmer switches, are connected to an equipment grounding conductor. A way to connect a metal faceplate to the equipment grounding conductor must be provided, even if a metal faceplate is not installed. These

Trade Tip

Approved switches are marked with the current and voltage or, if they are horsepower-rated, the maximum rating that they are designed for.

installation requirements make snap switches part of an effective ground-fault current path when either of the following conditions is met:

- The switch is mounted with metal screws to a metal box or metal cover that is connected to the equipment grounding conductor. A nonmetallic box with integral means for connecting to an equipment grounding conductor can also be used.
- An equipment grounding conductor or equipment bonding jumper is connected to an equipment grounding termination of the snap switch.

Surface-type snap switches that are used with open insulator wiring are required to be mounted on an insulating material that separates the conductors by at least 1/2 inch from the surface that is wired, per **NEC** 404.10. Flush-type snap switches can be mounted in boxes that are set back from the finished surface as long as the extension plaster ears are seated against the surface and with the mounting yoke or strap of the switch seated against the box. When in the off position, a switching device will completely disconnect all of the ungrounded conductors to the load the switch controls.

FIGURE 9.3

A general snap switch.

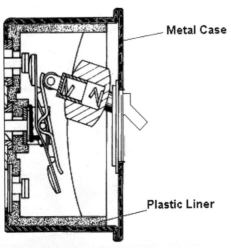

Metal Case

Plastic Liner

LUMINAIRES, LAMPHOLDERS, AND LAMPS

NEC 410 covers luminaires, lampholders, pendants, conductors, and any equipment associated with them. "Luminaire" is another name for "lighting fixture." In **NEC** Article 100, a luminaire is defined in detail as a complete lighting unit that consists of a light source, such as a lamp, combined with any parts that are designed to position the light source and connect it to a power supply. It may also include parts that are designed to protect the light source or the ballast or to help distribute the light.

Article 410 covers line-voltage lighting systems, which are typical 120-volt, 240-volt, and 277-volt light-fixture installations. A variety of luminaries are covered in **NEC** 410, such as lampholders, pendant fixtures, receptacles, rosettes, incandescent filament lamps, arc lamps, and electric-discharge lamps.

Luminaires, lampholders, and lamps cannot have any live parts exposed to contact. Any exposed accessible terminals in lampholders or switches cannot be installed in metal fixture canopies or in the open bases of portable or floor lamps. An exception is made for cleat-type lampholders, provided they are at least 8 feet above the floor, in **NEC** 410.5. Whenever light fixtures are used in wet or damp locations, they have to installed in such a way that water doesn't enter the fixture or accumulate in any wiring areas, compartments, lampholders, or other electrical parts. If you plan to install a light fixture in a damp or wet area, it must be rated for these locations and marked as "suitable for wet locations" or "suitable for damp locations," per **NEC** 410.10.

All luminaires and lampholders must be listed, which means that they must be included in an approved list published by an organization that is acceptable to the authority having jurisdiction. The listing will state that

---

CODE UPDATE

In the 2008 **NEC**, the definition of luminaire was revised to include parts that protect fixture components and distribute light. Additionally, this section now specifies that an individual lampholder is not a luminaire.

Did You Know?

The 2008 **NEC** expanded Article 410.160 to include decorative lighting fixtures, lighting accessories for temporary seasonal and holiday use, and portable flexible lighting.

the equipment, material, and any required or recommended service size meets established designated standards or that it has been tested and found suitable for a specified purpose. Any fixture that is not clearly listed should not be installed.

NEC 410.10(D) establishes which types of luminaires are not allowed to be installed within a zone of 3 feet horizontally and 8 feet vertically from the top of the bathtub rim or shower stall. Fixtures that are prohibited from installation in this bath zone are track lights, pull chains, cord-attached suspended lights, pendant fixtures, and paddle fans. Fixtures that are allowed in the bathtub or shower zone must be listed for damp locations or for wet locations if they are going to be subjected to shower spray. The 3-foot-by-8-foot bathtub and shower zone does not apply to recessed or surface-mounted luminaires, switches, or receptacles. For switch requirements refer to **NEC** 404.4; receptacles in bathrooms are covered in **NEC** 406.8(C).

Lights that are a part of or installed separately in commercial range hoods must meet all four of the following requirements, per **NEC** 410.10(C):

- They must identified for use in the commercial cooking hood and meet any temperature requirements of the fixture.

CODE UPDATE

Strings of lights enclosed in flexible plastic, often called "rope" lights, have a power supply cord with a fused attachment plug and are intended for outline and decorative lighting. They are now included in **NEC** 410.

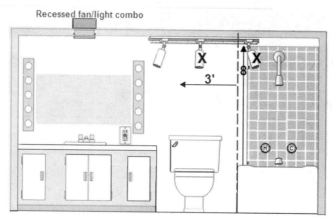

FIGURE 9.4

Track lighting cannot to be installed within the shower and bathtub zone. A recessed fan/light combination and vanity lights are allowed.

- The luminaires must be constructed in a listed manner so that exhaust, cooking vapors, grease, and oil cannot penetrate the lamp or wiring compartment. Any fixture diffusers must be thermal shock-resistant. Additionally, the fixture surface must be flat so that it does not collect deposits and is easy to clean.

- Any exposed parts of the luminaire must be protected against corrosion.

- None of the wiring methods or materials that supply the fixture can be exposed within the hood.

Luminaires for indoor sports areas, mixed-use, or "all-purpose" facilities must be protected from damage. **NEC** 410.4(E) requires that you use mercury-vapor or metal-halide lamps in playing and spectator seating areas in these locations. Luminaires of this type can have added guards for extra protection.

NEC 410.11 requires that if you are installing fixtures near any kind of combustible material, the luminaires you choose must include shades or guards that will keep the combustible material from being exposed to the temperatures in excess of 194 degrees F that could be generated by the heat from the luminaires. If the installation will be over combustible materials as opposed to just near them, then **NEC** 410.12 requires that they be

Trade Tip

The only kind of externally wired fixtures that are permitted in show windows are chain-supported luminaires.

unswitched. Unless an individual switch is installed for each fixture, lampholders have to be at least 8 feet above the floor or located and guarded in a way that prevents the fixture lamps from being removed or damaged.

Light Fixtures in Closets

When planning around safety concerns, it might surprise you that installing lighting in a closet presents enough challenges that **NEC** 410.16 is very specific about the requirements. Think about it. Closets can be filled with flammable materials such as cardboard boxes, paper, and cloth that may not necessarily be stored in a neat and organized manner. The last thing you want is for the high heat emitted by many types of light sources to cause damage to materials located in a closet or set them on fire. For just this reason, the **NEC** provides very detailed requirements; unfortunately, they are not necessarily easy to understand. Light-fixture installations in closets are driven by the proximity of the fixture to the surrounding areas in the closet. This is termed "storage space." The code refers to the volume bounded by the sides and back of the closet walls and a vertical plane from the floor. This length is measured from the floor up to the highest clothes hanging rod or 6 feet, whichever is higher.

For surface incandescent fixtures, the luminaire must have a clearance of 12 inches between it and the nearest point of storage and 24 inches between it and the back or side walls. If shelves installed in the closet are wider than 12 inches, the area above the shelves, regardless of the width of the shelves, is considered a storage area.

The figure below shows a particularly tricky closet-fixture installation, because the light is mounted on a sloped surface. It requires 6 feet from the floor to the closet rod, 24 inches from the side and back wall to the fixture, and 12 inches from the shelf to the fixture. When measuring 12 inches from the shelf to the fixture installation location, remember that the light must be 12 inches in front of the edge of the shelf *and* 12 inches above the shelf.

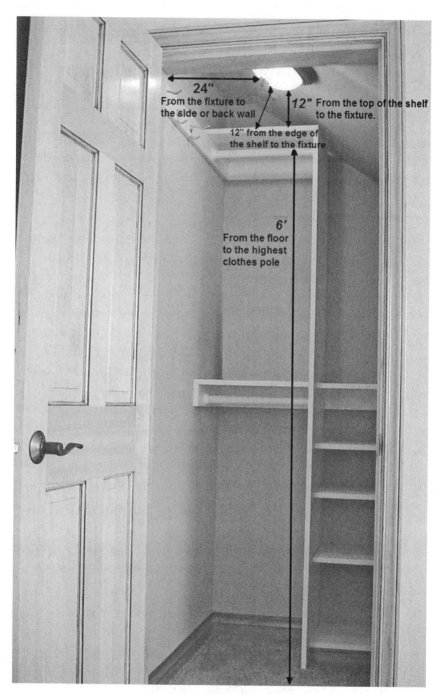

FIGURE 9.5

A closet installation.

Trade Tip

NEC 410.16 does not allow pendant lamp holders or incandescent fixtures with open or partially enclosed lamps to be installed in clothes closets.

The following types of luminaires are permitted to be installed in clothes closets:

- Surface-mounted or recessed incandescent fixtures designed with a completely enclosed lamp
- Surface-mounted or recessed fluorescent luminaires
- Surface-mounted LED (per **NEC** 410.16(A)3) or fluorescent fixtures that are identified as suitable for installation in a storage area

Clearance requirements for lighting fixtures from the uppermost shelf surface in a clothes closet are based on the type of fixture and are as follows:

- 12 inches for surface-mounted incandescent or LED luminaires with a completely enclosed light source installed on the wall above the door or on the ceiling
- 6 inches for surface-mounted fluorescent luminaires installed on the wall above the door or on the ceiling
- 6 inches for recessed incandescent or LED luminaires with a completely enclosed light source installed in the wall or the ceiling
- 6 inches for recessed fluorescent luminaires installed in the wall or the ceiling

Fixture Connections and Supports

Electric-discharge luminaires that are supported independently from an outlet box are required to be connected to the branch circuit through a metal raceway, nonmetallic raceway, Type MC cable, Type AC cable, Type MI cable, or nonmetallic sheathed cable, based on **NEC** 410.24. All light fixtures have to be securely supported, regardless of the cable type or installation location. If a luminaire weighs more than 6 pounds or is bigger than 16 inches in any dimension, it cannot be supported by just the screw shell of a lampholder, per **NEC** 410.30.

If a metal or nonmetallic pole is going to be installed to support luminaires and as a raceway to enclose supply conductors, then you have to meet the following requirements:

- If a pole is used, it must have a hand hole that is at least 2 inches × 4 inches with a cover that makes it suitable for use in wet locations. Access to the supply terminations within the pole or pole base must be provided.
- A threaded fitting or nipple that is brazed, welded, or attached to the pole opposite the hand hole for the supply connection should be used if raceway risers or cable is not installed within a pole.
- A metal pole must have an equipment grounding terminal that is accessible from the hand hole unless the pole is 8 feet or smaller and the luminaire can be removed for access.
- A metal pole must have any hinged base bonded.
- Metal raceways or other equipment grounding conductors have to be bonded to the metal pole with an equipment grounding conductor, per the requirements of **NEC** 250.118, and sized in accordance with **NEC** 2580.122.
- Conductors in vertical poles used as raceway have to be supported as outlined in **NEC** 300.19.

Trade Tip

You are not required to have a hand hole in a pole that is less than 8 feet high (above grade) if the supply wiring continues without any splices or pull points and the interior of the pole is accessible if the fixture is removed. If a pole is 20 feet or less in height, with a hinged base, then no hand hole is required.

Trade Tip

Other than 2-wire or multiwire branch circuits that supply power to luminaires that are connected together, branch-circuit wiring cannot pass through an outlet box that is an integral part of a fixture unless the luminaire is specifically identified for through-wiring.

NEC 410.21 requires that luminaires be constructed and installed so that the conductors in the outlet boxes are not be subjected to temperatures that are greater than what the conductors are rated for.

NEC 410.22 defines a completed installation as one in which each outlet box is provided with a cover unless already covered by means of a luminaire canopy, lampholder, receptacle, or similar device. Luminaires have to be securely fastened mechanically to ceiling framing by bolts, screws, or rivets. Fixture studs that are not a part of outlet boxes have to be made of steel, malleable iron, or similar materials. Any raceway fittings that are used to support fixtures have to be constructed in a way that they can support the weight of the complete fixture assembly and lamps. Luminaires are also allowed to be connected to busways, as long as the connection is in accordance with **NEC 368.17(C)**.

NEC 410.42 requires exposed metal parts of light fixtures to be connected to an equipment grounding conductor or insulated from the conductor and other conducting surfaces. These live parts cannot be accessible to unqualified personnel. Luminaires directly wired or attached to outlets that have wiring that doesn't provide a means of attaching to an equipment grounding conductor have to be made of insulating material and cannot have any exposed conductive parts. However, mounting screws, clips, lamp tie wires, and decorative bands on glass that are spaced at least 1 1/2 inches away from lamp terminals do not have to be grounded.

Did You Know?

Outdoor lighting fixtures and associated equipment can be supported by trees, per **NEC 410.22(G)**.

Light-Fixture Wiring Requirements

Wiring on or in luminaires has to be neatly arranged. Excess wiring has to be avoided, and none of the wiring can be exposed to physical damage, based on **NEC** 410.48. **NEC** 410.50 requires that conductors have to be arranged so that they are not subjected to temperatures above those that they are rated for. The screw shells of fixture lampholders need to be wired so that they are connected to the same luminaire, circuit conductor, or terminal, and the grounded conductor has to connected to the screw-shell lampholder.

Feeder and branch-circuit conductors that are within 3 inches of a ballast must have an insulation temperature rating that is not lower than 90 degrees C or 194 degrees F, as required by **NEC** 410.68. Additionally, you have to ensure that there is adequate airspace between lamps, shades, or other enclosures and combustible material.

All luminaires have to be marked with the maximum lamp wattage or electrical rating, manufacturer's name, or a similar method of identification. **NEC** 410.74 requires that if a light fixture requires supply wire that will be rated higher than 60 degrees C or 140 degrees F, the minimum supply-wire temperature rating has to be marked on the luminaire and the shipping container. A light-fixture electrical rating has to include the voltage and frequency and indicate the current rating of the unit, including the ballast, transformer, or autotransformer.

As described in **NEC** 410.78, tubing that is used for arms and stems that have cut threads cannot be any thinner than 0.040 inches, or 0.025 inches if it has rolled, pressed threads. If a metal canopy will support lampholders and shades that are heavier than 8 pounds or that have an attachment-plug receptacle, it cannot be any thinner than 0.020 inches. Other steel canopies cannot be less than 0.016 inch, and if they are made of other metals, they cannot be thinner than 0.020 inch. Pull-type canopy switches cannot be attached to the rims of metal canopies that are less than 0.025 inch thick, unless the rims are reinforced. Regardless of whether pull-type canopy switches are mounted in the rims or elsewhere in sheet-metal canopies, they cannot be located more than 3 1/2 inches away from the center of the canopy. These thickness requirements apply to measurements made on formed canopies.

NEC 410.82 requires portable luminaires to be wired with a flexible cord that complies with **NEC** 400.4 and an attachment plug that is polarized or grounded. Portable hand lamps have to abide by the additional requirements listed below:

- Metal shell, paper-lined lampholders cannot be used.

- Handlamps have to be equipped with a molded composition or insulating material handle.

- Handlamps must also have a substantial guard attached to the lampholder or handle.

- Metallic guards have to grounded by an equipment grounding conductor run with circuit conductors within the power-supply cord.

- Portable hand lamps are not required to be grounded if they are supplied through an isolating transformer with an ungrounded secondary that is not larger than 50 volts.

For double-pole switched lampholders, **NEC** 410.93 requires that the ungrounded conductors of a circuit and the switching device simultaneously disconnect both conductors of the circuit. **NEC** 410.102 requires that switched lampholders have a switching mechanism that interrupts the electrical connection to the center contact. The switching mechanism can also interrupt the electrical connection to the screw shell if the connection to the center contact is simultaneously interrupted.

An incandescent general-use lamp on a lighting branch circuit cannot have a medium base if it is rated over 300 watts or with a mogul base if it is rated over 1500 watts, as outlined in **NEC** 410.103.

Fast Fact

All light-fixture wiring has to be tested for short circuits and ground faults prior to being connected to the circuit.

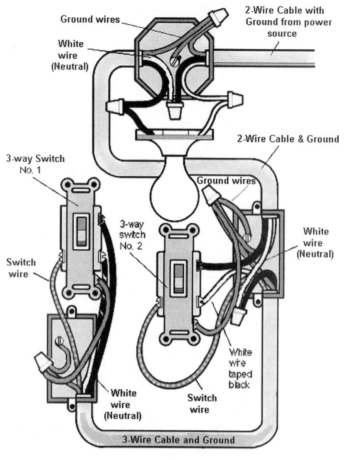

FIGURE 9.6

A complete three-way switch installation for a luminaire.

Trade Tip

Light fixtures cannot be installed adjacent to combustible material that could be subjected to temperatures in excess of 90 degrees C or 194 degrees F unless it is recessed in fire-resistant material in a fire-resistant construction. Even then, potential temperatures cannot exceed 150 degreesC or 302 degrees F.

FIGURE 9.7

*Thermal protection is not required in a recessed luminaire
installed in poured concrete.*

A recessed fixture that is not specifically identified for contact with insulation must have all recessed luminaire parts spaced at least 1/2 inch away from combustible materials, per **NEC** 410.116. Only the points of support and the trim finishing are permitted to make contact with combustible materials, unless it is a Type IC recessed fixture. Thermal insulation cannot be installed above a recessed luminaire or within 3 inches of the recessed luminaire's enclosure, wiring compartment, or ballast unless it is a Type IC fixture.

Fixture-Wiring Requirements

NEC 410.117 requires that branch-circuit conductors have a type of insulation that is suitable for the temperature where the fixture will be installed.

FIGURE 9.8

*Recessed light fixtures attached and spaced on framing
or other combustible material.*

These conductors are permitted to terminate in the luminaire. Tap conductors can be run from the luminaire terminal connection to an outlet box that is located at least 1 foot away from the light fixture. These tap conductors must be in a suitable raceway, Type AC, or Type MC cable that is a minimum of 18 inches and a maximum of 6 feet long.

Terminals for an electric-discharge lamp are considered energized if any lamp terminal is connected to a circuit of over 300 volts.

NEC 410.130(1) requires that the ballast and replacement ballasts of a fluorescent indoor fixture have integral thermal protection. A simple reactance ballast in a fluorescent luminaire with straight tubular lamps or a ballast in a fluorescent exit fixture does not have to be thermally protected. A ballast in a fluorescent luminaire that is used for egress lighting and is only energized during a power failure does not need thermal protection either.

CODE UPDATE

NEC 410.130(A) was revised in the 2008 edition to require that equipment for use with electric-discharge lighting systems and designed for an open-circuit voltage of 1000 volts or less be identified for this type of service.

> ### Did You Know?
>
> Luminaires with a metal-halide lamp, other than a thick-glass parabolic reflector lamp (PAR), must have a containment barrier that encloses the lamp or be provided with a physical means that only allows the use of a Type O lamp.

Recessed high-intensity luminaires installed in wall or ceiling cavities must have thermal protection and be identified as thermally protected unless the design, construction, and thermal performance characteristics are equivalent to a thermally protected fixture and identified as inherently protected.

Light-Fixture Disconnecting Means

Other than for dwellings, indoor fluorescent luminaires that use double-ended lamps and contain ballast(s) that can be serviced in place can have a disconnecting means that is either internal or external to each luminaire. Disconnecting means are not required under the following conditions:

- For luminaires installed in hazardous locations
- For emergency lighting
- In industrial establishments with restricted public access where maintenance and supervision will only be performed by qualified persons
- In environments where more than one luminaire is installed and supplied by a source other than a multiwire branch circuit, as long as the disconnecting means does not result in an illuminated space being left in total darkness

> ### Trade Tip
>
> The line side terminals of a luminaire's means of disconnecting must be guarded.

CODE UPDATE

In the 2008 edition of the **NEC**, Article 410.130(G) was revised to clarify the requirement for fluorescent luminaires that utilize double-ended lamps and contain ballasts. Now, if the disconnecting means is external to the light fixture, it must be a single device and must be located within sight of the luminaire.

Equipment Used with Luminaires

NEC 410.37 covers materials that are not part of a light fixture but that are used with them. For example, **NEC** 410.37(A) requires that auxiliary equipment, such as reactors, capacitors, and resistors, that are not installed as part of a luminaire assembly must be enclosed in accessible, permanently installed metal cabinets. However, separately mounted ballasts that are directly connected to a wiring system are not required to be separately enclosed. Wired luminaire sections are paired, with a ballast that supplies lamps in both. You can use #12 AWG FMC in lengths up to 25 feet to form an interconnection between paired fixture units.

NEC 410.138 requires that if an autotransformer is used to raise voltage for a fixture to more than 300 volts as part of a ballast for supplying lighting units, it can only be supplied by a grounded system. Equipment that has an open-circuit voltage over 1000 volts is prohibited from use in dwelling occupancies.

An external operable switch or circuit breaker that opens all ungrounded primary conductors is required for luminaires or lamp installations. The switch or circuit breaker has to be located within sight of the luminaires or lamps, unless it has a means of being locked in the open position. The ability to lock or add a lock to the disconnecting means has to remain in place at the switch or circuit breaker whether the lock is installed or not, which

Did You Know?

The terminal of an electric-discharge lamp is considered to be a live part.

CODE UPDATE

The 2008 edition of **NEC** 410.141(B) has been revised to require that locking means not be of a portable type that would be easily moved or relocated from the disconnecting means.

---

means that portable means for adding a lock to the switch or circuit breaker are not allowed.

NEC 410.143 deals with transformers for luminaires. A transformer secondary circuit voltage cannot exceed 15,000 volts, nominal, under any load condition, and the voltage to ground of any output terminals of the secondary circuit cannot exceed 7500 volts under any load conditions. Transformers need to have a secondary short-circuit current rating that is not more than 150 mA if the open-circuit voltage is over 7500 volts and not more than 300 mA if the open-circuit voltage rating is less than 7500. Additionally, these secondary circuit outputs can not be connected in parallel or in series. Transformers are required to be installed as close to the lamps as practicable in order to keep the secondary conductors as short as possible.

NEC 410.146 requires that if a luminaire or secondary circuit of tubing has an open-circuit voltage over 1000 volts, it has to be labeled with letters that are at least 1/4 inch high using terminology such as "Caution _____ volts." The voltage indicated in this labeling is the rated open-circuit voltage.

Track Lighting

NEC 410.151 provides the requirements for track lighting. Lighting track has to be installed in a permanent manner and permanently connected to a branch circuit. Only lighting-track fittings can be installed on lighting

---

CODE UPDATE

Article 410.151(B) of the 2008 **NEC** added an FPN that clarifies that the calculated load of the lighting track does not limit the length of track on a single branch circuit or the number of luminaires on a single track.

FIGURE 9.9

Track lighting is rated for a maximum number of heads based on voltage.

track, and lighting-track fittings cannot be equipped with general-purpose receptacles.

The connected load on lighting track cannot exceed the rating of the track and must be supplied by a branch circuit that is not rated for more than the rating of the track.

There are a number of locations where lighting track cannot be installed:

- Areas where the track is likely to be subjected to physical damage or corrosive vapors
- In wet or damp locations
- In storage-battery rooms or hazardous locations
- In concealed areas or areas where it extends through walls or partitions
- Less than 5 feet above a finished floor unless it is protected from physical damage or the track operates at less than 30 volts rms open-circuit voltage

Fittings identified for use on lighting track must be designed specifically for the track on which they are to be installed. They must be securely fastened to the track, maintain polarization and connections to the equipment grounding conductor, and be designed to be suspended directly from the track.

NEC 410.153 identifies heavy-duty lighting track as track identified for use over 20 amperes. Each fitting attached to a heavy-duty lighting track must have individual overcurrent protection. Lighting track has to be securely mounted so that a single section 4 feet or shorter in length has two supports. If the track is installed in a continuous row, each individual 4-foot section needs one additional support.

Lighting track also needs to be grounded in accordance with **NEC** 250, and the track sections need to be securely coupled to maintain continuity of the circuitry, polarization, and grounding throughout.

Because there are so many kinds of lighting conditions and types of fixtures, you can now see why this part of the **NEC** goes into so much detail. From here on, the installations we will be covering will be focused on specific types of electrical applications and will draw from the strong base of knowledge you have gained so far.

10

Motor and Branch-Circuit Overload Protection

WHAT YOU NEED TO KNOW

Up to this point in our analysis of the **NEC**, whenever we have examined overcurrent protection we have discussed using a circuit breaker that provides overcurrent, short-circuit, and ground-fault protection. However, motor overcurrent protection is achieved by separating overload-protection devices from short-circuit and ground-fault devices. Overload protection protects the motor as well as the motor control equipment and the branch-circuit conductors from motor overload and excessive heating. It does not provide short-circuit or ground-fault protection, which is provided by branch and feeder breakers. Essentially, you end up with two forms of protection; one that safeguards the motor and circuit conductors from the motor and another that provides ground-fault and short-circuit protection. While some motors have overload devices integrated into the motor starter, you can install a separate overload device, such as a dual-element fuse, for each ungrounded conductor at the load end of the branch circuit. In this case, you still have to install separate short-circuit and ground-fault protection. These installation arrangements, discussed in **NEC 430**, make motor calculations different from those used for other types of loads.

TERMS TO KNOW

Adjustable Speed Drive: The combination of a power converter, motor, and auxiliary devices used to control the speed of a motor with several pre-set ranges that make the speed "adjustable." Examples are thermal switches and heaters.

Controller: A switch or device normally used to start and stop a motor by making and breaking the circuit connection.

F.L.C.: The full-load current rating of a motor.

Horsepower: A unit of measuring power; 1 Horsepower = 33,000 foot-pounds of "work," or power exertion, per minute.

Locked-Rotor Kilovolt-Amps: The current, measured in kilovolt-amperes, drawn by a stalled electric motor.

Motor-Control Circuit: The circuit that carries electrical signals to a motor-control unit or system and directs the performance of the controller. Motor-control circuits do not carry main power current.

Motor Feeder: A feeder conductor for motors sized to the motor FLC rating, not the motor nameplate current rating.

Overload: An operating current that, if it continues for a sufficient length of time, would cause damage to the motor or motor circuits or generate dangerous overheating.

System Isolation Equipment: Remotely operated equipment that provides lockout capabilities in the motor-control circuit instead of in the power circuit.

Thermal Protection: A feature that protects a motor from overheating by temporarily cutting power to the motor for a cooling period.

Trip Current: The minimum value, or amount, of current that causes the circuit-breaker to trip without an intentional time delay.

MOTOR PROTECTION

Rating Motors

While many motor-overload devices are integrated into the motor starter, you can use a separate overload device such as a dual-element fuse, which is usually located near the motor starter, not the supply breaker. In order to pick the proper amperage for this device, you have to ascertain which factors to use to determine the external overload protection. There is no guesswork to this decision. NEC 430.6(A) dictates that for motors rated more than 1 hp without any integral thermal protection and for automatically started motors that are 1 hp or less, you must install an overload device that is sized based the motor nameplate full-load current rating. You have to size the overload devices no larger than the requirements of 430.32. The overload device is selected to trip, or be rated at, no more than the percentages of the motor nameplate full-load current rating.

NEC 430.6(1) refers you to Tables 430.247, 430.248, 430.249, and 430.250 because the full-load current rating of a motor is only used to determine the ampere rating of switches and the sizes of conductors, disconnects, and short-circuit and ground-fault protection. Don't use the current rating on the motor nameplate for this purpose. You will only use the actual nameplate current rating for motors that are built for reluctance-motor (RM) speeds less than 1200, high torques, or multi-speed motors. This bears repeating: do not use the current rating on the

FIGURE 10.1

Overload-device ratings.

SEPARATE OVERLOAD DEVICE RATING	
Motors with a marked service factor on the nameplate of 1.15 hp or greater	125%
Motors with a marked temperature rise on the nameplate of 40°C	125%
All other motors	115%

motor nameplate for determining conductor ampacity, the branch-circuit short-circuit ground-fault protection device, or the ampere rating of switches. Use the motor nameplate current ratings to select the devices to use to protect motors, their control apparatus, and branch-circuit conductors against excessive heat caused by motor overload and failure to start.

Connecting Motor Controls

You will need to connect motor-controller control-circuit devices and terminals with copper conductors. For torque motor-control conductors, use 14 or smaller AWG connected to a minimum of 7 pounds per inch for screw-type pressure terminals. NEC 430.14 requires you to locate motors in a manner that will provide adequate ventilation and facilitate maintenance. When dealing with motor protection, it's also critical to know how to properly connect motor terminals and where to locate motors within the system.

Based on **NEC** 430.22, you cannot size motor branch-circuit conductors smaller than 125 percent of the motor full-load-current (FLC) rating listed in Tables 430.147, 430.148, and 430.150, and you need to size the branch-circuit short-circuit and ground-fault protection device based on the requirements of both **NEC** 240.6(A) and 430.52(C)(1) Ex. 1. If you use fuses, you have to plan on one fuse for each ungrounded conductor, as required by **NEC** 430.36 and 430.55. In other words, a 3-phase motor would require three fuses. The fuses are installed at the load end of the branch circuit; remember that they don't provide short-circuit or ground-fault protection.

Fast Fact

When you are sizing conductors, you need to allow for a motor's inrush current. Inrush is the effect of drawing several times the motor's full load current while the motor is starting.

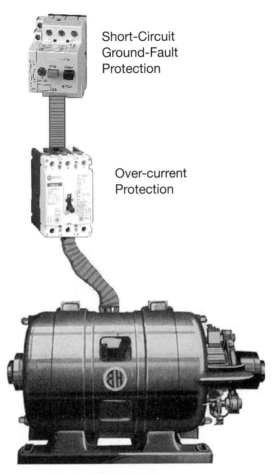

Short-Circuit
Ground-Fault
Protection

Over-current
Protection

FIGURE 10.2

Overcurrent protection is accomplished by separation from the short-circuit and ground-fault protection device.

Trade Tip

The **NEC** doe not require the wire size you use to address any possible voltage drop or the inability to start the motor that could result if the conductor is excessively long.

Demands and Characteristic of Different Types of Motors

The tables and standards in **NEC** Article 400 refer to various types of motor installations. It is important to know the diversity of these motors, because motors energize, pull, and utilize current in several different ways. Motor types are listed below:

- Continuous-duty motors are rated to be run continuously without any damage to or reduction in the life of the motor.

- Dual-voltage motors operate on two voltages so they can use two different power lines, one low and one high, such as 110 volts or 220 volts.

- Multispeed motors operate at variable speeds and have to be rated at the highest speed at which the motor can be started.

- Part-winding start-induction motors, also referred to as synchronous motors, are designed to start by initially energizing part of their primary armature winding, usually half, and then the rest in one or more steps.

- Torque motors utilize torque, the measurement of how much energy an engine develops each time it turns over. This energy is translated to power when you increase the RPMs. These motors are rated in revolutions per minute rather than in horsepower.

- Wye-start, delta-run motors are polyphase induction motors that utilize two or more phases of an alternating-current power line. A lower starting current demand minimizes the impact on the power system and reduces the chance of the motor tripping.

Conductor sizes and the number and location of motor-overload units and devices are determined by which of the power utilization methods a motor requires. Since motor loads have different characteristics than other loads, overcurrent protection for motors and motor circuits has unique requirements that differ from the rules for general loads that were specified in **NEC** Article 240. Motor circuits often demand a large amount of current initially to start up, usually around six times the normal full-load current of the motor. This inrush current is referred to in the **NEC** as the "locked

> ### Trade Tip
>
> When you are installing or inspecting a motor circuit, check these four attributes:
>
> - The branch-circuit (conductor) sizing
> - The presence of overload protection
> - The use of a branch-circuit short-circuit ground-fault protective device
> - The rating of the motor disconnect

rotor current." The actual full-load current for different motors of the same size and type varies, which is why **NEC** tables must be used for general motor applications. In this way, you ensure that if a motor fails and has to be replaced, the conductors and components of the motor circuit don't have to be replaced as well. The rules for torque motors and alternating-current adjustable-voltage motors are different. Based on the description of these motors and how they use current, you need to go by the actual nameplate current to size the components of these circuits.

The motor nameplate information is important. The nameplate voltage and horsepower ratings are needed in order to use the tables in Article 430. The horsepower rating at the applied voltage is used with the appropriate table to determine the full-load current rating of the motor. This full-load current value must be used to size the conductors and the branch-circuit short-circuit and ground-fault protective device.

BRANCH-CIRCUIT PROTECTION

Sizing Motor Branch-Circuit Conductors

Probably the best way to understand the methods for sizing motor branch-circuit conductors is to look at a few examples of the process. Here are the considerations for sizing motor conductors:

- The very first thing you have to do is determine the motor's full-load ampacity from the motor nameplate.

- Single motor branch-circuit conductors cannot have an ampacity lower than 125 percent of the full-load current for the motor as it is listed in **NEC** Tables 430.147, 430.148, 403.149, and 430.150.

- You have to refer to **NEC** Table 310.16 to select the branch-circuit conductor size from Table 310.16, based on the terminal temperature rating (60 or 75 degrees C) of the equipment.

Now we will put the process to work in a couple of examples. In our first, let's figure out what size THHN copper conductor you would need for a 2-hp, 230V, single-phase motor:

- Since you have to size the conductor sized at no less than 125 percent of motor FLC and you don't know the FLC, refer to **NEC** Table 430.148, which shows the FLC of 2-hp, 230V, single-phase motor as being 12 amperes.

- Multiply 12 amps × 1.25 (125 percent), which equals 15 amperes.

- Refer to **NEC** Table 310.16 in the column for type THHN copper, and you will see that **14 AWG** THHN is rated for 20 amps at 60 degrees C.

Let's try an example with a little less information provided up front. Let's say you have a continuous-duty, 115V, 1 1/2-hp electric motor with a nameplate rating that indicates that it draws 18.5 amperes. However, when you check Table 430.148, 20 amperes are required for the motor, and, based on the standards in **NEC** 430.6a, you have to go by the size listed in the table. The next factor you have to consider is that, based on **NEC** 430.22(a), a single motor used in a continuous duty situation (defined as running non-stop for three hours or more) must have an ampacity that is not less than 125 percent of the motor full-load current. Here you will take the nameplate FLC of 18.6 amperes and multiply it by 125 percent. The result is 23.5 amperes, which is higher than the 20 amperes you obtained from Table 430.148. This means that the minimum conductor size for the motor circuit, after any ampacity adjustments or corrections have been applied, needs to be at least 23.5 amperes.

NEC Table 310.16, allows you to use #12 THWN copper conductors for an installation like this one and permits some cable assemblies, such as Type NM, to be sized at #12 for this application. Ultimately, there are a number of other factors that can affect the size of a motor branch-circuit conductor, such as a voltage drop that can result in long conductor ampacity adjustment factors due to the number of current-carrying conductors in the same raceway or adjustments for ambient temperature of the installation.

Let's look at one more example, but this time we are going to calculate the branch-circuit conductor size for a 3-phase motor. It is a 7.5-hp, 230V motor with conductor terminals that are rated at 60 degrees C. This calculation breaks down into two quick steps:

- Look at **NEC** Table 430.150, and you'll find that a 7.5-hp 230V, 3-phase motor has an FLC of 22 amperes. You know that you have to size the branch-circuit conductor at no less than 125 percent of FLC, so multiply 22 × 1.25 and you come up with 27.5 amperes.
- From **NEC** Table 310.16 you'll see that a #10 AWG conductor is rated for 30 amps at 60 degrees C.

Sizing Motor Branch-Circuit Short-Circuit and Ground-Fault Protection

The **NEC** requires that you size motor branch-circuit short-circuit and ground-fault protective devices based on the values provided in Table 430.148. **NEC** 430.51 pre-empts the general short-circuit and ground-fault protection requirements that are listed in **NEC** Article 240. Additionally, the protective devices that you use for motors cannot exceed the value you calculate using the percentage values given in **NEC** Table 430.52. These percentages are based on the type of fuse or circuit breaker you plan on using and are included in your calculations to allow the motor to be started without causing the protection device to trip on the locked rotor, starting current. The percentages are:

- Dual-element time-delay fuse: 175 percent
- Nontime-delay fuse: 300 percent

Trade Tip

The short-circuit and ground-fault protection devices that are required for motor circuits are not the same types required for dwelling units (**NEC** 210.8), feeders (**NEC** 215.9 and 240.13), or services (**NEC** 230.95)

- Instantaneous-trip circuit breaker: 800 percent
- Inverse-time circuit breaker: 250 percent

If the results of calculating these percentages don't equal the standard size of a fuse or nonadjustable circuit breaker or the available settings of an adjustable circuit breaker, then you need to use the next standard size. Ultimately you are seeking to ensure that the short-circuit and ground-fault protection will be large enough to handle the motor's inrush current.

Let's say you were going to use a noontime-delay fuse as the short-circuit ground-fault protective device for a 115V, 1 1/2-hp motor. The fuse size would have to be calculated using the value of 18.6 from **NEC** Table 430.148, which would then be multiplied by 300 percent. The resulting size would be 55.8, which you would round up to the next higher standard size as permitted by **NEC** 430.52(c)(1) Ex. No. 1. This means that the short-circuit ground-fault protective device, in this case the noontime-delay fuse, would be 60 amperes.

Now you might think that a fuse this size would not be adequate to protect the #12 copper conductor that is allowed for the branch-circuit conductor size. But remember that you aren't going to protect the motor branch circuit based on the ampacity rules of **NEC** 240. There is an additional level of protection provided to motor circuits, and that is overload-protection devices.

Did You Know?

Short-circuit and ground-fault protection devices are designed for fast current rise, short-duration events, while overload-protection devices are designed for slow current increases and long-duration scenarios.

If overload relays are provided, branch-circuit, short-circuit and ground fault protection, per **NEC** 430.52 and Table 430.152, is provided by Class RK5 (FRN-R/FRS-R), RK1 (LPN-RK/LPS-RK), and J (LPJ) fuses at 125, 130, and 150 percent of motor full-load current, respectively, or by the next higher standard size if these do not correspond to a standard fuse size.

Without overload relays, where the fuse is the only motor-overload protection, Class RK5 (FRN-R/FRS-R) fuses provide overload, branch-circuit, short-circuit, and ground-fault protection based on **NEC** 430.32 and must meet the following criteria:

- If the motor has a 1.15 service factor or 40-degrees C rise, size the fuse at 110 to 125 percent of the motor nameplate full-load current.

- For motors smaller than a 1.15 service factor or over 40-degrees C rise, size the fuse at 100 to 115 percent of the motor nameplate full-load current.

OVERLOAD PROTECTION OF MOTORS AND BRANCH CIRCUITS

Overload-protection devices safeguard motors, motor controls, and motor branch-circuit conductors against any excessive heating that might result from a motor overload and failure to start. Overload protection alone doesn't protect against short circuits or ground faults, but the combination of an overload-protective device and the branch-circuit short-circuit ground-fault protective device result in overcurrent protection for the motor and motor circuits. Sometimes, motor-overload protection exists in the motor itself. If so, the motor will be labeled as "thermally protected," or with the abbreviation "TP." When a motor does not already have integral thermal protection, you have to provide external overload protection in the form of fuses or breakers.

NEC Section 430.32 (A) requires that each continuous-duty motor that is rated at more than 1 horsepower has be protected by an overload-protective device rated at no more than the code-specified percentages of the motor nameplate rating.

Modifications to these values are allowed if the resulting size would not be sufficient to start the motor or carry the motor load, per **NEC** 430.34. Motor overload-protective devices usually aren't designed to open for short

SEPARATE OVERLOAD DEVICE RATING	
Motors with a marked service factor on the nameplate of 1.15 hp or greater	125%
Motors with a marked temperature rise on the nameplate of 40°C	125%
All other motors	115%

FIGURE 10.3

Rate-overload protection requirements.

circuits or ground faults, so they have to be protected by fuses or circuit breakers that meet the rating requirements of **NEC** 430.52. Often, motor starters and controllers that have thermal overload devices will specify a maximum rating for fuses or circuit breakers required to protect the internal overload device from damage. The maximum ampere rating for overload-protection devices is provided in **NEC** Table 430.72(B). Find the control-circuit conductor size you are going to use, then identify the type of protection to be provided, such as separate protection, and then match the type of wire, such as copper, to determine the adequate amperage. You *must* pay attention to the notes below the table in order to determine many of the overcurrent-protection device sizes. Comprehensive overcurrent protection for a motor, the motor branch circuit, and motor controls is achieved through combining a properly sized motor branch-circuit short-circuit and ground-fault protective device, such as fuses, circuit breakers, or motor circuit protectors, with an adequate overload- protection device. For a multispeed motors, each winding connection must be considered separately. **NEC** 430.55 permits a single branch-circuit, short-circuit, and ground-fault protective device to provide the combined required protection if the branch-circuit short-circuit and ground-fault protective device is set or rated to also provide overload protection in accordance with the requirements of **NEC** Section 430.32 or 430.34. You can look at Example #D8 in Appendix D of the code as a reference for motor overcurrent-protection requirements.

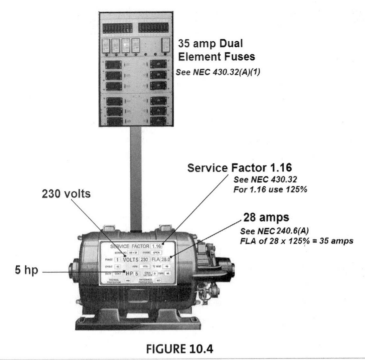

35 amp Dual
Element Fuses
See NEC 430.32(A)(1)

Service Factor 1.16
See NEC 430.32
For 1.16 use 125%

230 volts

28 amps
See NEC 240.6(A)
FLA of 28 x 125% = 35 amps

5 hp

FIGURE 10.4

Determine overload-protection device size from the service factor rating of 1.15, the voltage of 230, and the motor size of 5 hp.

Motor Disconnects and Controllers

Rating the disconnecting means for a general motor installation is done in accordance with Part VI of **NEC** Article 430. The disconnecting means has to be able to disconnect the motor and controller from the circuit. **NEC** 430.74 states that motor-control circuits have to be arranged so that they will disconnect from all power sources when the disconnect is in the open position. The disconnect can be comprised of two or more separate devices, at least one of which disconnects power to the motor and controllers and the others disconnect the current to the motor-control circuits. If you use separate devices, they have to be installed immediately adjacent to each other. The motor controller is used to start and stop a motor by actually breaking the motor-circuit current.

The motor controller is required to open only as many conductors of the circuit as necessary to start and stop the motor. For example, one conductor must open to control a 2-wire, single-phase motor and two conductors must open to control a 3-wire, 3-phase motor. Each motor requires its own controller, per **NEC** 430.87, and an enclosure that is suitable for the environment that the controller is installed in, based on **NEC** Table 430.91. A molded case switch, rated in amperes, can serve as a motor controller. For stationary motors rated at 2 hp and 300V or less, the controller can be either of the following:

- A general-use switch with an ampere rating that is no less than two times the full-load current rating of the motor
- A general-use snap switch with a motor full-load current rating that does not exceed 80 percent of the ampere rating of the switch

A motor starter is a form of motor controller and thus is a contactor with a proper horsepower rating. Ratings of the controller or motor starter must be in accordance with **NEC** VII. Each controller must be capable of starting and stopping the motor it controls and have the capability of interrupting the locked-rotor current of the motor, per **NEC** 430.82. Rating a controller is outlined in **NEC** 430.83 and includes considerations such as horsepower, the amperage of any branch-circuit inverse-time circuit breakers, case switches, motor size, and types of motors. Manual switches or circuit breakers that are rated in horsepower can be used as a disconnecting means and as the controller for motor circuits. **NEC** 430.81(A)(1) allows for stationary motors that

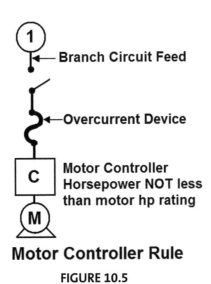

Motor Controller Rule

FIGURE 10.5

The motor controller has less horsepower than the motor.

are smaller than 1/8 horsepower and normally left running, such as clock motors, to use the branch circuit as a motor controller.

If you are going to wire for a portable motor installation and the motor is only 1/3 horsepower or smaller, you can use a receptacle as the motor-control device, per **NEC** 430.81(B).

FIGURE 10.6

Branch circuit used as a motor controller.

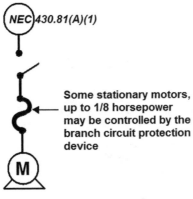

Rule Exception

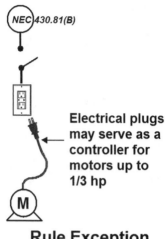

Electrical plugs may serve as a controller for motors up to 1/3 hp

Rule Exception

FIGURE 10.7

Attachment plug and receptacle used as the motor controller.

Rating of motor controllers must comply with **NEC** 430.83. Controllers must have horsepower ratings that are at an application voltage no lower than the motor's horsepower. Additionally, you can use a branch-circuit inverse-time circuit breaker that is rated in amperes as a controller for any motor, per **NEC** 430.83(A)(2).

FIGURE 10.8

Branch-circuit inverse-time circuit breakers used as a motor controller.

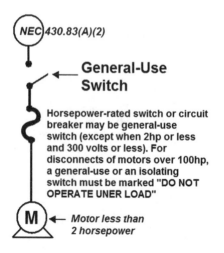

General-Use Switch

Horsepower-rated switch or circuit breaker may be general-use switch (except when 2hp or less and 300 volts or less). For disconnects of motors over 100hp, a general-use or an isolating switch must be marked "DO NOT OPERATE UNER LOAD"

Motor less than 2 horsepower

FULL LOAD CURRENT AND OTHER DATA FOR THREE PHASE A.C. MOTORS						
Motor Horsepower	Motor Ampere	Size Breaker	Size Starter	Heater Ampere	Size Wire	Size Conduit
1	230V 4.2	15	00	4.830	12	3/4"
	460 V 2.1	15	00	2.415	12	3/4"
1-1/2	230V 6.0	15	00	6.900	12	3/4"
	460V 3.0	15	00	3.450	12	3/4"
2	230V 6.8	15	0	7.820	12	3/4"
	460V 3.4	15	00	3.910	12	3/4"
3	230V 9.6	20	0	11.040	12	3/4"
	460V 4.8	15	0	5.520	12	3/4"
5	230V 15.2	30	1	17.480	12	3/4"
	460V 7.6	15	0	8.740	12	3/4"
7-1/2	230V 22	45	1	25.300	10	3/4"
	460V 11	20	1	12.650	12	3/4"
10	230V 28	60	2	32.200	10	3/4"
	460V 14	30	1	16.100	12	3/4"
15	230V 42	70	2	48.300	6	1"
	460V 21	40	2	24.150	10	3/4"
20	230V 54	100	3	62.100	4	1"
	460V 27	50	2	31.050	10	3/4"
25	230V 68	100	3	78.200	4	1-$1/2$"
	460V 34	50	2	39.100	8	1"

1) Overcurrent device may have to be increased due to starting current and load conditions.
   See 2008 NEC 430.250, Table 430-250. Wire based on 75°C terminations and insulation
2) Overload heater must be based on motor nameplate and sized per NEC 430-32.
3) Conduit size based on Rigid metal conduit with some spare capacity. For minimum size and other conduit types, see NEC Appendix C

FIGURE 10.9

Cross-reference table for motors.

For motor circuits that are 600 volts or less, you have to install the controller's manual disconnecting means within sight of the motor controller and no more than 50 feet away, with two exceptions. The term "within sight" means visible. The controller disconnect can be out of sight of the controller as long as the controller is marked with a warning label giving the location and identification of the disconnecting means and the disconnecting means can be locked in the open position. Each motor and controller or magnetic starter must have some form of approved manual disconnecting means, rated in horsepower, or a circuit breaker. When this disconnecting means is in the open position, it must disconnect both the controller and the motor from all ungrounded supply conductors and must clearly indicate whether it is in the open or the closed position.

You can easily become confused when attempting to size all the elements that are required for motor installations, unless you take each aspect one step at a time. Lay it all out on paper if need be so that you can be sure to include all the variables required to meet the **NEC** standards for motors.

Refrigeration and Air Conditioning

WHAT YOU NEED TO KNOW

NEC Article 440 focuses on electric motor-driven air-conditioning and re-frigerating equipment and the branch circuits and controllers needed for them. There is a large section devoted to the special considerations required for circuits that supply hermetic refrigerant motor-compressors and the air-conditioning and refrigerating equipment that is supplied from a branch circuit that supplies hermetic refrigerant motor-compressors. The main considerations are the operating amperes of the motors, over-current protection, circuit sizing, and ground-fault protection. The basic knowledge you gained in the previous chapters concerning these applications will be put to use any time you have to install refrigeration or air-conditioning units.

TERMS TO KNOW

Branch-Circuit Selection Current. The value in amperes used instead of the rated-load current to determine the required ratings for motor branch-circuit conductors, disconnecting means, controllers, and branch-circuit, short-circuit, and ground-fault protective devices. Wherever the running overload-protective device allows a sustained current, the value that is greater than the percentage of the rated-load current. The value of branch-circuit selection current is always equal to or greater than the marked rated-load current.

Hermetic Refrigerant Motor Compressor. A combination of equipment that consists of a compressor and motor that are enclosed in the same housing as the motor operating in the refrigerant unit without any external shaft or shaft seals.

Leakage-Current Detector Interrupter (LCDI). A device installed in a power-supply cord that senses any leakage current flow occurring between or from the cord conductors and then interrupts the circuit at a predetermined level.

Rated-Load Current. For a hermetic refrigerant motor-compressor, the current that flows when a motor compressor kicks on at the rated load, rated voltage, and rated frequency of the equipment it serves.

HERMETIC REFRIGERANT MOTOR-COMPRESSOR IDENTIFICATION

A hermetic refrigerant motor compressor has a nameplate that identifies the name of the manufacturer, any identifying designations, and the phase, voltage, and frequency of the compressor. This information is required by **NEC** 440.4. The rated-load current in amperes of the motor compressor must be listed by manufacturer on the motor-compressor nameplate and/or the nameplate of the equipment in which the motor compressor is used. Additionally, you will need to locate the locked-rotor current of each single-phase motor compressor that has a rated-load current of more than 9 amperes at 115 volts or more than 4 1/2 amperes at 230 volts and the quantity of polyphase motor-compressors on the motor-compressor nameplate. If a thermal protector complying, as outlined in **NEC** 440.52(A)(2) and (B)(2) is used, then the motor-compressor or equipment nameplate must be labeled as "thermally protected."

Multimotor and combination-load equipment must have a similar visible nameplate labeled with the manufacturer's name, the rating in volts, and

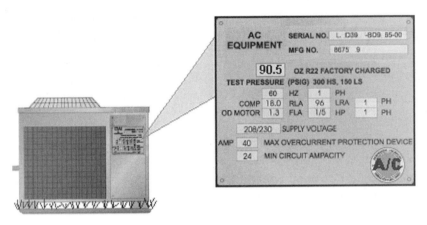

FIGURE 11.1

Nameplate on hermetic refrigerant motor-compressor equipment.

the frequency and number of phases. Additionally, the minimum size of the supply-circuit conductor ampacity, the maximum rating of the branch-circuit short-circuit and ground-fault protective device, and the short-circuit current rating of the motor controllers or industrial control panel must be determined. The ampacity is calculated using Part IV of this article and is based on the total number of motors and other loads that will be operated at the same time. The branch-circuit short-circuit and ground-fault protective-device rating is calculated using Part III and cannot exceed the values derived. For multimotor and combination-load equipment that will be used on two or more circuits, the required labeling has to be provided for each circuit.

Room air conditioners that comply with **NEC** 440.62(A) do not have to be marked with the minimum supply-circuit conductor ampacity or the maximum rating of the branch-circuit short-circuit and ground-fault

Trade Tip

Multimotor and combination-load equipment that is connected to a single 15- or 20 ampere, 120- or 240-volt, single-phase branch circuit is marked as a single load.

Did You Know?

If there isn't a rated-load current listed on an equipment nameplate, the rated-load current on the compressor nameplate is used.

protective device. You will also find that multimotor and combination-load equipment that is installed using cord-and-attachment-plug-connected equipment and equipment supplied from a branch circuit protected at 60 amperes or less in single and two-family dwellings do not have to be marked with a short-circuit current rating.

The branch-circuit selection current must be listed on the nameplate of any hermetic refrigerant motor compressor or equipment that contains one of these compressors. Any protection system that is approved for use with the motor compressor and protects it while allowing continuous current that may be in excess of the specified percentage of nameplate rated-load current provided in **NEC** 440.52(B)(2) or (B)(4) has to be marked with a branch-circuit selection current that complies with that section of the code.

MULTIPLE MOTOR INSTALLATIONS

If you are installing multiple motors, they must comply with **NEC** 440.7 as well as 430.24, 430.53(B) and 430.53(C), and 430.62(A). The highest-rated, largest motor is assumed to be the motor with the highest rated-load current. If two or more motors have the same highest rated-load current, you would only consider one of them as the highest-rated motor. The full-load current that you need to use to determine the highest-rated motor is derived from the corresponding motor horsepower rating listed in **NEC** Tables 430.247, 430.248, 430.249, and 230.258; however, hermetic refrigerant motor compressors and fan or blower motors are covered by **NEC** 440.6(B)

DISCONNECTING METHODS

Under the provisions of **NEC** 430.87, an air-conditioning or refrigerating system is considered to be a single machine. The motors are allowed to be remotely located from each other.

Single-Phase Alternating-Current Motors: Full-Load Currents in Amperes

For motors running at usual speeds and motors with normal torque characteristics
Voltages listed are rated motor voltages
Currents are permitted for system voltage ranges of 110 - 120 and 220 - 240 volts

Horsepower	115 Volts	200 Volts	208 Volts	230 Volts
1/6	4.4	2.5	2.4	2.2
1/4	5.8	3.3	3.2	2.9
1/3	7.2	4.1	4.0	3.6
1/2	9.8	5.6	5.4	4.9
3/4	13.8	7.9	7.6	6.9
1	16	9.2	8.8	8.0
1½	20	11.5	11.0	10
2	24	13.8	13.2	12
3	34	19.6	18.7	17
5	56	32.2	30.8	28
7½	80	46.0	44.0	40
10	100	57.5	55.0	50

FIGURE 11.2

*Horsepower and voltage ratings from **NEC** 430.248.*

NEC 440.12 requires that you select the disconnecting means that serves a hermetic refrigerant motor compressor based on the nameplate rated-load current or branch-circuit selection current, whichever is greater, and the locked-rotor current of the motor compressor.

The amperage rating must be at least 115 percent of the nameplate rated-load current or branch-circuit selection current, whichever is greater. You can use a listed unfused motor circuit switch without fuse-holders with a horsepower rating that is not less than the equivalent horsepower described in **NEC** 440.12(A)(2) for determining equivalent horsepower. It can have an ampere rating that is less than 115 percent of the specified current.

To determine the equivalent horsepower that complies with **NEC** 430.109, choose the horsepower rating from **NEC** Tables 430.248 through 430.250 that corresponds to the rated-load current or branch-circuit selection current, whichever is greater, and also the horsepower rating from **NEC** Table 430.251(A) or 430.251(B) that corresponds to the locked-rotor current.

When the combined load of two or more motors is simultaneous on a single disconnecting means, the rating for the disconnect is determined using **NEC** 440.12(B)(1) and (B)(2). This is true whether you are using two or more hermetic refrigerant motor compressors or one or more, hermetic refrigerant motor compressors with other motors.

To arrive at the required horsepower rating for the disconnecting means, you need to add together all the currents, including resistance loads, at the rated-load condition, and be sure to include the locked-rotor condition. The combined rated-load current and the combined locked-rotor current are considered to be a single motor for this section of **NEC** 440.12(B)(1)a and (B)(1)b.

NEC 440.12(B)(1)(a) requires that you choose the full-load current equivalent to the horsepower rating of each motor, except for a hermetic refrigerant motor compressor or a fan or blower motor, from **NEC** Tables 430.248 through 430.250. Add these full-load currents to the motor-compressor rated-load currents or branch-circuit selection currents, whichever is greater, and then to the rating in amperes of other loads. This will give you an equivalent full-load current for the combined load.

NEC 440.12(B)(1)(b) refers you to **NEC** Tables 430.251(A) and (B) to get the locked-rotor current equivalent for the horsepower rating of each motor. For fan and blower motors that are either shaded-pole or permanent

---

Trade Tip

If the nameplate rated-load current, branch-circuit selection curren,t and locked-rotor current do not correspond to the currents shown in Table 430.248, Table 430.249, Table 430.250, Table 430.251(A), or Table 430.251(B), the next higher horsepower rating value can be used. If you arrive at different horsepower ratings when you use these tables, a horsepower rating that is at least equal to the largest of the values you obtain must be used.

> **Trade Tip**
>
> If part of the concurrent load is a resistance load and the disconnecting means is a switch that is rated in horsepower and amperes, then the switch needs to have a horsepower rating that is not less than the combined load to the motor compressors and other motors at the locked-rotor condition. The ampere rating of the switch cannot be less than this locked-rotor load plus the resistance load.

split-capacitor types marked with the locked-rotor current, you will use the marked value. This rule does not apply to hermetic refrigerant motor compressors. The locked-rotor currents are added to the motor-compressor locked-rotor currents and to the rating in amperes of other loads. This gives you an equivalent locked-rotor current for the combined load. In some installations of two or more motors or loads, such as resistance heaters, the motors cannot be started simultaneously, so you have to use combinations of locked-rotor and rated-load current or branch-circuit selection currents to determine the equivalent locked-rotor current for the simultaneous combined load. In this case, you would use whichever of these currents is larger as the equivalent locked-rotor current for the simultaneous combined load.

The ampere rating of the disconnecting means has to be at least 115 percent of the sum of all motor currents at the rated load, based on **NEC** 440.12(B)(1), unless the disconnect is an unfused motor-circuit switch without fuseholders that has a horsepower rating not less than the horsepower listed in **NEC** 440.12(B). In this case, the disconnect can have an ampere rating that is less than 115 percent of all the currents. For small motor-compressors that may not have a locked-rotor current marked on the nameplate or that are not covered in **NEC** Tables 430.247-250, you have to assume that the locked-rotor current is six times the rated-load current.

If the rated-load or locked-rotor current for the equipment you will be connecting requires a disconnecting means that is rated at higher than 100 hp, then you have to comply with the requirements of **NEC** 430.109(E). Remember that early on in this book we told you that you would have to look forward and backward during the course of using the code. As you can

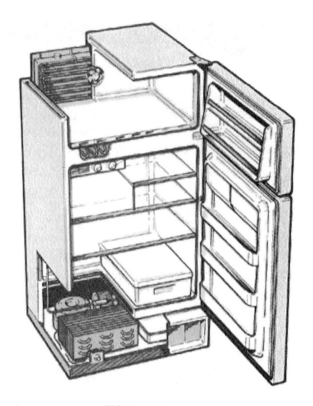

FIGURE 11.3

A refrigerator uses a cord and receptacle for the disconnecting means.

see in this chapter, you have to refer back to Article 430 quite often for refrigerant and motor installations.

Cord-connected equipment such as soda machines, room air conditioners, household refrigerators, and the like that have a separate connector or a plug that attaches to a receptacle can consider unplugging from the outlet as an approved disconnecting means.

CODE UPDATE

NEC 440.15 in the 2008 edition prohibits a disconnecting means from blocking the equipment's nameplate.

Did You Know?

Do you know the difference between accessible and readily accessible? In the case of equipment, something is considered accessible if it provides close approach and is not limited due to its elevation or by locked doors. It is readily accessible if it can be reached quickly without requiring you to climb over or remove obstacles.

Disconnecting means have to be located so that they are within sight of the air-conditioning or refrigerating equipment they are associated with and must be readily accessible.

There are two exceptions to **NEC** 430.102(A). The first is for refrigerating or air-conditioning equipment that is essential to an industrial process in a facility. As long as only qualified persons can service this equipment, you can use a disconnecting means that can be locked in the open position. If you add a locking disconnect, it has to be installed on or at the switch or circuit breaker and must remain in place with or without the lock installed. The second exception is for installations using an attachment plug and receptacle as the disconnecting means in accordance with **NEC** 440.13. This disconnect must be accessible but is not be required to be readily accessible.

OVERCURRENT PROTECTION

Part III of **NEC** Article 440 covers the standards of overcurrent protection from short circuits and ground faults for branch-circuit conductors, control devices, and motors in circuits associated with hermetic refrigerant motor compressors.

Motor-compressor branch-circuit short-circuit and ground-fault protection devices have to be able to carry the starting current of the motor. You are required to use a protective device with a rating or setting that does not exceed 175 percent of the motor-compressor rated-load current or branch-circuit selection current, whichever is greater. However, if this calculation results in a device size that would not be sufficient to handle the starting current of the motor, you have are allowed to increase the device

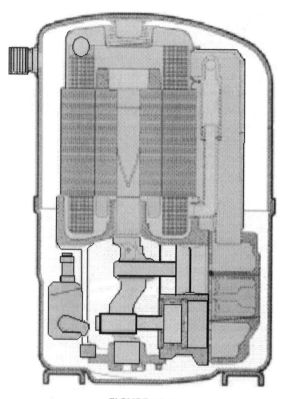

FIGURE 11.4

Hermetic refrigerant motor.

size, as long as it doesn't exceed 225 percent of the motor rated-load current or branch-circuit selection current, whichever is greater. In any of these cases, you should not use a branch-circuit, short-circuit or ground-fault protective device that is smaller than 15 amperes.

Any equipment branch-circuit short-circuit and ground-fault protective devices must be capable of carrying the starting current of the equipment. If a hermetic refrigerant motor compressor is the only load on the circuit, this protection needs to comply with **NEC** 440.22(A). However, if the equipment has more than one hermetic refrigerant motor compressor or includes a hermetic refrigerant motor compressor and other motors or equipment that draw current, then short-circuit and ground-fault protection is governed by **NEC** 430.53 and 440.22(B)(1) and (B)(2).

In a situation such as this, if the hermetic refrigerant motor compressor is the largest load connected to the circuit, then the branch-circuit short-circuit and ground-fault protective-device rating cannot exceed the values specified in **NEC** 440.22(A) for the largest motor compressor plus the sum of the rated-load currents or the branch-circuit selection current (whichever is greater) of the other motor compressors and other loads. However, if the hermetic refrigerant motor compressor is not the largest load connected to the circuit, then the rating of the branch-circuit short-circuit and ground-fault protective device cannot be larger than the sum of the rated-load current or branch-circuit selection current rating (whichever is greater) for the motor compressors plus the value specified in **NEC** 430.53(C)(4). If only non-motor loads are supplied in addition to the extra compressors, then the total of rated-load currents or branch-circuit selection-current ratings cannot be greater than the value listed in **NEC** 240.4. The exceptions to this standard are as follows:

- Equipment that starts and operates on a 15- or 20-ampere 120-volt or a 15-ampere 208- or 240-volt single-phase branch circuit can be protected by a 15- or 20-ampere overcurrent device that protects the branch circuit. However, if the maximum branch-circuit short-circuit and ground-fault protective-device rating marked on the equipment is less than 15 or 20 amperes, the circuit protection device cannot exceed the value that is marked on the equipment nameplate.

- The nameplate marking for cord-and-plug-connected equipment such as household refrigerators, soda machines, or water coolers that are not rated any larger than 250 volts, should use single-phase to determine the proper branch-circuit size. Each one of these units is considered to be a single motor unless the nameplate is marked otherwise.

Fast Fact

The protective-device rating cannot exceed the manufacturer's values marked on the equipment.

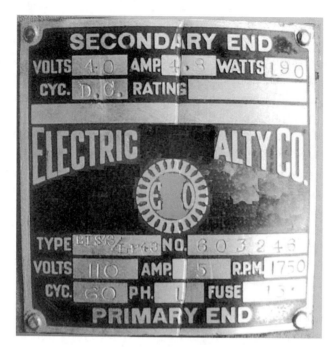

FIGURE 11.5

The nameplate of this hermetic refrigerant motor indicates 15-ampere protection.

CONDUCTOR AMPACITIES

The provisions in **NEC** 440.31 do not apply to internal motor conductors or motor controllers or to conductors that form an integral part of the equipment. Branch-circuit conductors that supply one motor compressor cannot have an ampacity that is less than 125 percent of either the motor-compressor rated-load current or the branch-circuit selection current, whichever is greater. **NEC** 440.32 allows a wye-start, delta-run-connected

---

Trade Tip

The individual motor-circuit conductors for a wye-start, delta-run-connected motor compressor carry 58 percent of the rated load current. This is why a multiplier of 72 percent is used; it is the equivalent of multiplying 58 percent by 1.25 (125 percent).

motor-compressor to have branch-circuit conductors between the controller and the motor compressor that are sized at 72 percent of either the motor-compressor rated-load current or the branch-circuit selection current, whichever is greater.

If you have to size conductors to supply one or more motor compressors with or without an additional load, **NEC** 440.33 requires that the conductor ampacity cannot be less than the *sum* of the rated-load or branch-circuit selection-current ratings (whichever is larger) of all the motor compressors *plus* the full-load currents of the other motors *plus* 25 percent of the highest motor or motor-compressor rating in the group. There are two exceptions to this rule. The first is when the circuitry is interlocked in a way that prevents the second motor compressor or group of motor compressors from starting and running. In this case, the conductor size is based on the largest motor compressor or group of motor compressors that will be operating at a given time. The second exception indicates that branch-circuit conductors for room air conditioners are to be based on Part VII of Article 440.

COMBINATION LOAD

If you have a combination load, such as conductors that will be supplying a motor-compressor load and a lighting or appliance load, then **NEC** 440.34 requires that the conductors have an adequate ampacity to supply the lighting or appliance load *plus* the required ampacity for the motor-compressor load. Again, however, if the circuitry is interlocked in a way that prevents the motor compressor from simultaneously operating at the same time as the lighting or appliance loads, then the conductor size is based on the largest size required for the motor compressor and other loads to be operated at a given time.

The motor compressor, motor controller, and branch-circuit conductors have to be protected against excessive heating that could result if the motor were to overload or fail to start. **NEC** 440.52(A) outlines the approved methods for providing this protection as follows:

- A separate overload relay can be provided that responds to the motor-compressor's current and will trip at no more than 140 percent of the motor-compressor rated-load current.

- An integral thermal protector in the motor compressor that will prevent dangerous overheating due to overload and failure to start is also acceptable. If the current-interrupting device is separate from the motor compressor and its control circuit is operated by a protective device integral to the motor compressor, the design has to be arranged so that opening the control circuit will interrupt the flow of current to the motor compressor.

- A fuse or inverse-time circuit breaker is permitted and can also be used as the branch-circuit short-circuit and ground-fault protective device. This device needs to be rated larger than 125 percent of the motor-compressor rated-load current and must have a time delay that will allow the motor compressor to start and accelerate its load. The equipment for the motor compressor has to be marked with this maximum branch-circuit fuse or inverse-time circuit-breaker rating.

If an overload relay or other devices for motor overload protection are not capable of opening short circuits, they have to be protected by fuses or inverse-time circuit breakers with ratings or settings that comply with **NEC** 440 Part III, unless they are identified for group installation or part-winding motors. In this case the maximum size of fuses or inverse-time circuit breaker must be based on the motor nameplate rating.

NEC 440.55 indicates that overload protection for motor compressors and equipment that is cord-and-attachment-plug-connected used on a either a 15- or 20-ampere 120-volt or 15-ampere 208- or 240-volt, single-phase branch circuit can be used as outlined in **NEC** 440.55(A), (B), and (C).

The rating of the attachment plug and receptacle cannot exceed 20 amperes for 125 volts or 15 amperes for 250 volts, and the short-circuit and ground-fault protective device that protects the branch circuit must have a sufficient time delay to permit the motor compressor and other motors to start and accelerate their loads.

CODE UPDATE

The 2008 **NEC** revised Article 440.53 to require overload relays or other devices to be identified for group installations.

CODE UPDATE

Revisions to the 2008 **NEC** Article 440.55(A) require that both the controller and overload-protective device be identified for installation with the branch-circuit short-circuit and ground-fault protective device.

A cord-and-attachment-plug-connected room air conditioner is considered as a single motor unit when you are determining its branch-circuit requirements as long as it is not rated more than 40 amperes and 250 volts, single-phase. Additionally, the rating of the branch-circuit short-circuit and ground-fault protective device cannot exceed the ampacity of the branch-circuit conductors or the rating of the receptacle the unit will attach to, whichever is less. The total marked rating of the unit can exceed 80 percent of the rating of a branch circuit if no other loads are supplied or 50 percent if lighting outlets, other appliances, or general-use receptacles are also supplied.

An attachment plug and receptacle is an approved disconnecting means for a single-phase room air conditioner that is rated 250 volts or less as long as the manual controls on the unit are readily accessible and located within 6 feet of the floor or if a manually operated disconnecting means is installed in a readily accessible location within sight of the room air conditioner. If a flexible supply cord is used to connect a room air conditioner, the cord cannot be longer than 10 feet for a nominal 120-volt rating or 6 feet for a nominal 208- or 240-volt system.

NEC 440.65 requires single-phase cord-and-plug-connected room air conditioners to have a factory-installed LCDI (leakage-current detector-interrupter) or arc-fault circuit-interrupter (AFCI) protection that is either an integral part of the attachment plug or located in the power-supply cord within 12 inches of the attachment plug.

PUTTING IT ALL TOGETHER

This section of the code is very math-intensive. Let's look at a typical 230-volt, 4-hp air-conditioning unit that has a motor compressor and an additional load from the fan motor. The equipment nameplate indicates that the fan motor full-load current is 1.3 amperes, the motor is 18 amperes, and the motor compressor LRA (locked-rotor amperes) is 96 amperes. The conductors that you choose to supply this equipment must have an ampacity that is not less than 125 percent of *either* the rated-load or the branch-circuit selection current, whichever is larger, *plus* the full-load current of the fan motor:

$$(RLA \times 1.25) + \text{fan motor FLA} = \text{minimum supply circuit amapcity}$$
$$18.0 \times 1.25 + 1.3 = 23.8 \text{ amperes}$$

Round this ampacity up to 24 and compare it to the minimum ampacity of the branch-circuit conductors selected from **NEC** Table 310.16. This table shows a #12 Type TW, THW, or THWN copper conductor safely carrying 25 amperes continuously in an atmosphere where equipment is operated in an ambient temperature up to 86 degrees F. Remember that if the installation is in an attic or on a roof where it would exceed this ampacity, you would have to apply the correction factors listed in the table to increase the size of the conductors to compensate for the rise in temperature.

NEC Section 240.3(D) prohibits the overcurrent protection from exceeding 20 amperes for a #12 conductor unless otherwise specifically permitted in Sections 240.3(E) through 240.3(G), which instruct you to protect air-conditioning circuit conductors in accordance with Parts C and F of Article 440. So, now that we have consulted **NEC** 310 and 240, only to return to Article 440, let's look at an example of why we made the trip.

Let's look at a commercial refrigeration cooler, where the nameplate indicates a maximum fuse or circuit-breaker size of 40 amperes and a supply-circuit ampacity of 24 amperes. No derating for ambient temperature is required. But do you realize that a #12 copper conductor, which has an

---

Did You Know?

A short circuit occurs when ungrounded (hot) conductors fault together line-to-line. A ground fault results from ungrounded conductors fault to the equipment grounding conductor or grounded equipment.

ampacity of 25 amperes, is acceptable to supply this unit? The reason that a #12 AWG wire with an ampacity of 25 amperes is permitted to have over-current protection of 40 amps is because a 40-ampere fuse or circuit breaker installed at the origin of the circuit protects the conductors from short circuit and ground fault. The conductor is protected from overload by the overcurrent device that is contained in the motor controller. The combination of these two forms of protection provides the overcurrent protection necessary to meet code requirements for safety.

Now we will look at determining the size of a disconnection means for our air-conditioner example. The unit had a compressor-motor RLA of 18 amperes, which has to be added to the 1.3-ampere FLA of the fan motor. The total of 19.3 amperes is considered the equivalent full-load current for the combined load. Based on **NEC** Table 430.148, the full-load current rating of a 230-volt, single-phase, 3 horsepower motor is 17 amperes, and the full-load current rating of a 230-volt, single-phase, 5-horsepower motor is 28 amperes. Since the equivalent full-load current of our air-conditioner is 19.3 amperes, we have to size the disconnect switch based on the require-ments for a 5-horsepower, 230-volt, single-phase unit. Additionally, the rating of the disconnecting means has to be a least 115 percent of the sum of all of the currents at the rated-load condition:

$$115 \text{ percent} \times 19.3 \text{ amperes} = 22.3 \text{ amperes}$$

If the disconnecting means serves as the branch-circuit overcurrent protec-tion for the unit, then the overcurrent device size becomes the determin-ing factor in sizing the disconnecting means. An unfused disconnect switch based on this 115 percent rating and the horsepower rating would establish the minimum switch rating.

There is one other consideration in determining the correct size of the dis-connecting means serving our conditioning unit. The disconnect-means rating also needs to be based on the motor currents at locked-rotor condi-tion. **NEC** Table 430.151(A) provides the conversion of locked-rotor cur-rent to horsepower. In our example, the AC nameplate indicates that the motor-compressor LRA is 96 amperes. Since the nameplate does not give a LRA for the fan motor, we have to calculate it based on six times the FLA:

$$6 \times 1.3 \text{ amperes} = 7.8 \text{ amperes}$$

Add this to the motor-compressor LRA of 96 amperes and we get an equivalent LRA for the combined load of 103.8 amperes. If we compare this to **NEC** Table 430.151, we'll see that a single-phase, 230-volt motor with a 103.8-ampere motor locked-rotor current requires a disconnect switch that is based on the same value as a 5-horsepower rating.

CHAPTER

12

Generators and Transformers

WHAT YOU NEED TO KNOW

In order to size and install generators (**NEC** 445) and transformers (**NEC** 450), you need to have a basic understanding of induction. Induction is the process of generating electrical current in a conductor by placing the conductor in a changing magnetic field. It is called induction because the current is said to be "induced" by the magnetic field. Induction that occurs in an electrical circuit, affecting the flow of electricity by creating a change in current that results in a change in voltage in that same circuit, is called inductance.

A generator is essentially a backwards electric motor, because each time the motor spins it sends an exact amount of electricity through its connecting wires to an electric motor on the other end. A transformer changes or "transforms" energy by either increasing or decreasing the voltage that flows to a conductor. A transformer has a ferromagnetic core, usually made of iron, nickel, or cobalt and metals that have a high magnetic saturation point. This core is wrapped in multiple coils or "windings" of wire.

The input line connects to the primary coil, while the output lines connect to secondary coils. The alternating current in the primary coil induces a constant movement of alternating magnetic energy that flows around the core and then changes direction and induces an alternating current in each of the secondary coils.

The bottom line is that there are a lot of changes in energy taking place when either of these piece of equipment are utilized, so you have to know how to connect to them properly to provide an installation that is both safe and effective.

TERMS TO KNOW

AC Generator Exciter: An auxiliary generator that is used to provide field current for a larger generator or alternator.

Autotransformer: An electrical transformer that only has one winding. The winding has at least three electrical connection points, called taps. The voltage source is applied to two taps and the load is connected to two taps; one of those is usually a common connection that is also connected to the power source. In an autotransformer, a portion of the same winding effectively acts as part of both the primary and the secondary winding.

Balancer Set: A device that equalizes energy or transfers it from the less loaded side of a piece of equipment to the more heavily loaded side.

Generator: An engine that converts mechanical energy into electrical energy by electromagnetic induction

Transformer: A device used to increase or decrease the voltage in a conductor. It either steps up or steps down the voltage being transmitted.

Transformer Fault Sensing: A system within a transformer that senses a fault and reacts by opening a main switch or common-trip overcurrent device for a 3-phase, 4-wire system to protect against single-phasing or internal faults.

Vault: An enclosure for the live parts and conductors of a piece of equipment.

GENERATORS

As with other motors we have discussed, **NEC** 445.11 requires a generator to have a nameplate giving the manufacturer's name, rated frequency, power factor, number of AC phases, the subtransient and transient impedances, the rating in kilowatts or kilovolt amperes, a rating for the normal volts and amps, rated revolutions per minute, insulation-system class, any rated ambient temperature or temperature rise, and a time rating. The size and type of overcurrent protection devices will be based on this critical data.

NEC 445.12 defines the basic overcurrent-protection standards for various types of generators. A constant-voltage generator has to be protected from overloads by either the generator's inherent design or circuit breakers, fuses, or other forms of overcurrent protection considered suitable for the conditions of use. This is true except for AC generator exciters.

Two-wire, DC generators are only allowed to have overcurrent protection in one conductor if the overcurrent device is triggered by the entire current that is generated other than the current in the shunt field. For this reason,

FIGURE 12.1

All generators must have some form of overcurrent protection.

the overcurrent device cannot open the shunt field. If the generator operates at 65 volts or less and is driven by an individual motor, then the overcurrent protection device needs to kick in if the generator is delivering up to 150 percent of its full-load rated current. When a 2-wire DC generator is used in conjunction with balancer sets, it accomplishes the neutral points for the 3-wire system. This means it requires an overcurrent device that is sized to disconnect the 3-wire system if an extreme unbalance occurs in the voltage or current.

For 3-wire DC generators, regardless of whether they are compound or shunt-wound, one overcurrent device must be installed in each armature lead and connected so that it is activated by the entire current from the armature. These overcurrent devices need to have either a double-pole, double-coil circuit breaker or a 4-pole circuit breaker connected in both the main and equalizer leads plus two more overcurrent devices, one in each armature lead. The overcurrent protective devices have to be interlocked so that no single pole can be opened without simultaneously disconnecting both the leads of the armature from the system.

FIGURE 12.2

A 4-Pole circuit breaker is required for overcurrent protection of a 3-wire DC generator.

Fast Fact

If a generator is vital to the operation of an electrical system and the generator's failure would cause great personal risk, then the overload sensing device can be connected to an annunciator or alarm that would alert authorized personnel instead of interrupting the generator circuit.

Generator Conductors

The ampacity of the conductors that run from the generator terminals to the first distribution device that contains overcurrent protection cannot be less than 115 percent of the nameplate current rating for the generator, per **NEC** 445.13. An exception is made if the design and operation of the generator prevent overloading, in which case the ampacity of the conductors has to be equal to at least 100 percent of the nameplate current rating of the generator. You can size the neutral conductors based on **NEC** 220.61, but the conductors have to able to carry ground-fault currents that are not smaller than those required by **NEC** 250.30(A). Neutral conductors of DC generators that will carry ground-fault currents cannot be smaller than the minimum size of the largest conductor. For generators that operate at more than 50 volts to ground, any live parts have to be protected so that unqualified people don't accidentally come in contact with them.

Generator terminal housings have to comply with **NEC** 430.12. If you need a horsepower rating to determine the required minimum size of the generator terminal housing, compare the full-load current of the generator with comparable motors in **NEC** Tables 430.247 through 430.250. The higher horsepower rating in these tables should be used if the generator selection falls between two listed ratings.

CODE UPDATE

The 2008 edition of the **NEC** revised Article 445.13 to require sizing generator conductors that carry ground-fault current by using **NEC** 250.30(A) instead of **NEC** 250.24(C).

CODE UPDATE

The phase "lockable in the open position" was added to the 2008 version of **NEC** 445.18.

Generator Disconnects

According to **NEC** 445.18, all generators have to be equipped with at least one disconnect that is lockable in the open position and allows the generator and all its associated protective devices and controls to be disconnected entirely from the circuits that are supplied by the generator. The only exceptions to this rule are installations where both of the following conditions exist:

- The driving means for the generator can be readily shut down.
- The generator is not set up to operate in parallel with another generator or another voltage source.

A new section concerning generators that supply multiple loads was added to the 2008 **NEC** as 445.19. It permits a single generator that is installed to supply more than one load, or multiple generators that operate in parallel, as long as either of the following conditions are present:

- A vertical switchboard with separate sections.
- Individual enclosures with overcurrent protection are tapped from a single feeder for load separation and distribution.

Did You Know?

Connections for temporary power generators for dwellings, as in the case of a loss of power during extreme weather conditions, must be made by a licensed electrical contractor. Many homeowner's insurance companies will not otherwise cover any damages caused by running a portable generator.

Fast Fact

The calculation of the relationship between the primary and secondary voltage is: the primary coil voltage divided by the number of turns in the primary equals the total of the secondary coil voltage divided by the number of turns in the wire of the secondary:

$$\frac{\text{Primary coil voltage}}{\text{\# of turns of wire in the primary}} = \frac{\text{Secondary coil voltage}}{\text{\# of turns of wire in the secondary}}$$

TRANSFORMERS

NEC 450 covers the installation requirements for transformers. Transformers have a primary and a secondary voltage. Electricity moving through the primary coils of a transformer creates a magnetic field. This magnetic field induces an electric field and moves electrons in the secondary coil, which produces an electric current.

Let's say you know that the incoming, or "primary," voltage to a transformer is 120 volts. You also know that the primary has 75 turns and the secondary has 150 turns, but you don't know what the secondary voltage will be. All you have to do is start by dividing the primary volts by the number of primary turns: $120 \div 75 = 1.6$.

Trade Tip

Four kinds of transformers are exempt from the requirements of **NEC** 450:

- Current transformers
- Dry-type transformers that are a component part of other systems
- Transformers that are an integral part of an X-ray, high-frequency, or electrostatic-coating equipment
- Transformers used with Class 2 and Class 3 circuits that comply with **NEC** 725

Did You Know?

Nonlinear loads can increase heat in a transformer without causing the overcurrent protective device to activate.

Since you know that number of turns in the secondary, you would now multiply 150 secondary turns times the primary current of 1.6: 150 × 1.6 = 240.

Now you have determined that the secondary voltage is 240 volts. With this basic understanding of the relationship between voltages and turns in a transformer, you are ready to get a firm grip on the standards that regulate transformer installations.

Transformer Overcurrent Protection

Transformer overcurrent protection is governed by **NEC** 450.3(A) for transformers operating over 600 Volts, nominal; by 450.3(B) for transformers 600 volts, nominal or less; and by 450.3(C) for transformers installed as a motor-control circuit transformer. This type of voltage

FIGURE 12.3

Transformer overcurrent-protection methods are based on the voltage size of the transformer.

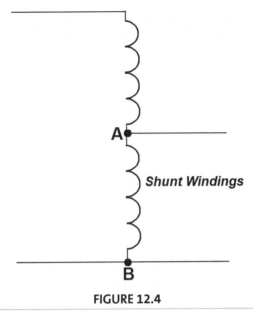

FIGURE 12.4

Overcurrent devices cannot be installed in series with the shunt winding between points A and B.

transformer is installed indoors or in an enclosed area and must be protected by primary fuses.

Article 450.3 assumes that you are installing either one single-phase transformer or a polyphasic bank of two or more single-phase transformers that will be operating as a unit.

Each autotransformer that is 600 volts, nominal, or less has to be protected by an individual overcurrent device installed in series, with each ungrounded input conductor to comply with **NEC** 450.4.

The overcurrent device has to be rated or set at no more than 125 percent of the rated full-load input current of the autotransformer. If your calculation results in a rated input current of 9 amperes or more but the size doesn't correspond to the standard rating of a fuse or nonadjustable circuit breaker, you need to use the next higher standard-size overcurrent device. If the rated input current of the autotransformer is less than 9 amperes, an overcurrent device rated or set at not more than 167 percent of the input current should be used. Overcurrent devices cannot be installed in series with the shunt winding of the autotransformer between Points A and B.

Fast Fact

The shunt winding is the winding that is common to both the input and the output circuits.

Grounding Autotransformers

NEC 450.5 provides the standards for grounding zigzag or T-connected autotransformers that are connected to 3-phase, 3-wire ungrounded systems in order to create a 3-phase, 4-wire distribution system or provide a neutral point for grounding. These transformers have a continuous per-phase current rating and a continuous neutral current rating. Autotransformers that are zigzag-connected cannot be installed on the load side of any system grounding connection, not even the connections specified in NEC 250.24(B), 250.30(A)(1), or 250.32(B). A grounding autotransformer that is used to create a 3-phase, 4-wire distribution system from a 3-phase, 3-wire ungrounded system must meet the following four requirements:

- It must be directly connected to the ungrounded phase conductors. The autotransformer may not be switched or provided with overcurrent protection that is independent of the main switch and common-trip overcurrent protection for the 3-phase, 4-wire system.

- There must be an overcurrent sensing device associated with the autotransformer that will cause the main switch or common-trip overcurrent protection to open if the load reaches or exceeds 125 percent of the transformer's continuous current per-phase or neutral rating. You may also provide delayed tripping for temporary overcurrents that would be sensed at the autotransformer overcurrent device. This would accommodate proper operation of the branch or feeder protection devices on the 4-wire system.

- You must install a fault-sensing system that will cause a main switch or common-trip overcurrent device for the 3-phase, 4-wire system to open and provide protection against single-phasing or internal faults.

> **Trade Tip**
>
> A fault-sensing system can be created by using two subtractive-connected doughnut-type current transformers to sense and produce a signal if an unbalance occurs in the line current to the autotransformer, which is 50 percent or more of the rated current.

- The autotransformer must have a continuous neutral-current rating that is large enough to handle the maximum *possible* neutral unbalanced-load current of the 4-wire system.

Grounding autotransformers used in high-impedance grounded neutral systems requires overcurrent protection to guard against a specified ground-fault current and must meet the following requirements of **NEC** 450.5(B):

- A continuous neutral-current rating adequate to handle the specified ground-fault current.

- An overcurrent protection device in the grounding autotransformer branch circuit that has an interrupting rating that complies with **NEC** 110.9 and will open all the ungrounded conductors simultaneously.

- Overcurrent protection that is set or rated for currents that don't exceed 125 percent of the autotransformer continuous per-phase current rating or 42 percent of the continuous-current rating of any series-connected devices in the autotransformer neutral connection. Any delayed tripping for temporary overcurrents in order to ensure the proper operation of ground-responsive tripping devices on the main system cannot exceed a value that would be more than the short-time current

> **Fast Fact**
>
> The phase current in a grounding autotransformer is one-third the neutral current.

rating of the grounding autotransformer or any series connected devices in the neutral connection to the transformer. However, high-impedance grounded systems with a maximum ground-fault current that is designed for up to 10 amps only for grounding autotransformers with grounded impedance and is rated for continuous duty can have an overcurrent device that is rated up to 20 amperes. This requires that it will simultaneously open all the ungrounded conductors andcan only be installed on the line side of the grounding autotransformer.

Transformer Secondary Ties

A secondary tie is a circuit that operates at 600 volts, nominal, or less between phases and connects two power sources or power-supply points, such as the secondaries of two transformers. **NEC** 450.6 explains that a secondary ties consists of one or more conductors per phase or neutral. Tie circuits require overcurrent protection at each end in compliance with **NEC** Parts I, II, and VIII of Article 240; however, **NEC** 450.6 provides alternate circuit-sizing options if overcurrent protection does not meet the requirements of **NEC** 240. For example, if all the loads are connected to transformer supply points at *each end* of a tie and the overcurrent protection does not meet **NEC** 240, then the tie's rated ampacity can be *67 percent* of the rated secondary current of the highest-rated transformer supplying the secondary tie system. If the load is connected to the *tie at any point between transformer supply points* and adequate overcurrent protection is not provided, the rated ampacity of the tie shall can be *no less than 100 percent* of the rated secondary current of the highest-rated transformer supplying the secondary tie system. Both of the supply ends of each ungrounded tie conductor must be equipped with a overcurrent protective device that will open when the tie conductor reaches a predetermined temperature under short-circuit conditions. This protection must either be what is commonly referred to as a limiter, which is a fusible-link cable connector, terminal, or lug sized to correspond with the conductor based on the operating voltage and the type of insulation on the tie conductors; or automatic circuit breakers that are activated by devices that have comparable time-current characteristics. Only when the tie circuits are made up of multiple conductors per phase can they be sized and protected based solefy on **NEC** 450.6(A)(4).

NEC 450.6(A)(4) describes the requirements for interconnection of phase conductors between transformer supply points. If the tie consists of more than one conductor per phase or neutral, the conductors of each phase or neutral have to comply with one of the following provisions:

■ The conductors are interconnected to create a load supply point, and one of the protective devices listed as in the preceding paragraph is provided in each ungrounded tie conductor at this interconnected point on both sides of connection. The ampacity of this interconnection can-not less than the load to be served.

■ The loads are connected to one or more individual conductors of a paralleled conductor tie without interconnecting the conductors of each phase or neutral and without the type of protection specified in either requirement above at load connection points. Under these conditions, the tie conductors of each phase or neutral must have a combined capacity ampacity that is not less than 133 percent of the rated secondary current of the highest-rated transformer that supplies the secondary tie system. The total load of the taps cannot be greater than the secondary current rating of the highest-rated transformer, and the loads must be equally divided on each phase and on the individual conductors of each phase.

Let's assume the operating voltage exceeds 150 volts to ground. In this case, if you use secondary ties with limiters, they need to have a switch at each end; when the switch opens, it must deenergize the tie conductors and limiters. The current rating of this switch has to be at least the same as the rated current ampacity of the conductors that are connected to the switch. The switch has to be made in such a way that it won't open if it is exposed to the magnetic forces that could result from short-circuiting current.

Trade Tip

Conductors that connect the secondaries of transformers as outlined in **NEC** 450.7 are not considered secondary ties.

An overcurrent device that is rated or set at not more than 250 percent of the rated secondary current of transformers with secondary ties must be provided in the secondary connections of each transformer that supplies the tie system. The secondary connection of each transformer must also have an automatic circuit breaker that is activated by a reverse-current relay. When a secondary tie system is grounded, each transformer secondary that supplies the tie system has to be grounded based on the requirements of NEC 250.30 for separately derived systems.

Physical and Mechanical Protection

NEC 450.8 begins the section on transformer mechanical protection. The purpose is to reduce the risk of damage to transformers from external causes. The first step is to install transformers inside a moisture-resistant enclosure that is not combustible and will help keep foreign objects away from the components.

All the energized parts have to be guarded, and warning signs or markings that list the operating voltage of the equipment must be placed on the equipment or its enclosure. The enclosure space also has to have adequate ventilation to control temperature rises that could occur from running the transformer. Any fences or exposed non-current-carrying metal parts of the transformer enclosure have to be grounded.

FIGURE 12.5

Transformers' distinguishing features and requirements must be taken into consideration before they are installed.

Transformer Nameplate

Information includes: kilovolt-amperes, frequency, primary & secondary voltages, impedance, insulating liquids

> ### Trade Tip
>
> Fire-resistant construction must have a minimum fire rating of 1 hour.

The nameplate on a transformer has to include the following: the name of the manufacturer, equipment rating in kilovolt-amperes, frequency, primary and secondary voltage, any required clearances from the transformer, and indication of any insulating liquid is used in the transformer, along with the amount and kind. If the transformer is 25 kVA or larger, than the unit's impedance must also be listed. All transformers and transformer vaults, except for dry-type transformers that are up to 600 volts, nominal, and located in the open, must be readily accessible to qualified personnel to provide for safe inspection and maintenance.

NEC 450.21 requires that dry-type transformers that are rated at 1121/2 kVA or less and are installed indoors have a minimum of 12 inches of clearance between the equipment and any combustible material unless a fire-resistant, heat-insulated barrier is installed. The same type of transformers that are rated over 112 1/2 kVA must be installed in a fire-resistant transformer room.

NEC 450.22 provides the standards for dry-type transformers that are installed outdoors. These locations require that the transformer enclosure be weatherproof. If the transformer is larger than 112 1/2 kVA, it can't be located within 12 inches of any combustible materials.

Transformers insulated with a dielectric nonflammable fluid can be installed indoors or outdoors. If the transformer is to be used indoors and is

> ### Trade Tip
>
> Nonflammable dielectric fluid that is referenced in **NEC** 450 does not have a flash point or fire point and is not flammable in air.

rated over 35,000 volts, it must be installed in a vault that has a liquid confinement area and a pressure-relief vent. The transformer itself must contain a means for absorbing any gases generated by arcing inside the tank, or the pressure-relief vent must be connected to a chimney or flue that can carry the gases away.

Oil-Insulated Transformers Installed Outdoors

In the case of oil-insulated transformers that are installed on roofs, attached to a building, or adjacent to combustible materials, a fire hazard exists that must be guarded against by using one or more of the following safety precautions:

- Space separations
- Fire-resistant barriers
- Automatic fire-suppression systems
- Enclosures that confine oil in the event of a ruptured transformer tank

Vault locations and requirements are an important consideration when installing a transformer and are affected by the size and type of transformer. Vaults must be located where they can be ventilated to the outside air without using flues or ducts wherever such an arrangement is practicable.

FIGURE 12.6

A fire-rated transformer vault encloses the transformer and restricts access to only qualified personnel.

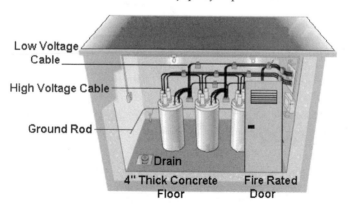

Fast Fact

Materials must not be stored in transformer vaults.

For transformer installations that require that the equipment be placed in a vault, the minimum fire-resistance rating for the walls and floors of the vault is 3 hours. Concrete floors of outside vaults have to be at least 4 inches thick. Vaults with concrete floors that have a vacant space or other stories below it must have a minimum fire resistance of 3 hours unless the transformer is protected by automatic sprinklers, water spray, carbon dioxide, or halon, in which case a 1-hour rating is permitted. **NEC** 450.43 requires transformer vault doors to be tight-fitting and have a minimum fire rating of 3 hours unless the vault is protected by similar fire protection-systems. In this case, the door can be rated at 1 hour. Any ventilation openings need to be located as far away from the door as possible.

If a vault has natural ventilation to an outdoor area, the combined net area of all ventilating openings, not counting any screens, gratings, or louvers, cannot be less than 3 square inches for every kVA of transformer capacity. It can never be less than 1 square foot for any transformer that is smaller than 50 kVA.

NEC 450.46 syates that wherever it is practicable, vaults that house transformers that are larger 100 kVA need to have a drain or some other way to carry away any accumulation of oil or water in the vault, that the floor must be pitched toward this drain.

Water pipes and duct system that are not associated with the transformer installation maynot enter or pass through a transformer vault. This does not apply to piping provided for vault fire protection or transformer cooling.

Although you may not have an overwhelming number of times when you need to install a generator or transformer, it is vital that you understand how they work and what is required to install them in a manner that provides maximum safety.

13

NEC Tables, Annexes, and Examples

In some ways, Chapter 9 of the **NEC** is like a quick reference library. In the tables in the beginning of the chapter you'll find listings of the radius of conduits and tubing bends, the dimensions and percentage of area for various types of conduit, approved dimensions of insulated conductors and fixture wire, and even conductor size requirements based on direct-current resistance. One important thing to remember about the annexes is that they are not part of the code requirements; rather they are included for your information and reference.

TABLE AND TIPS

Here is some extra information that clarifies some of the Tables in **NEC** Chapter 9:

- Table 1 only applies to complete conduit or tubing systems. It is not meant to apply to sections of conduit or tubing that are used to protect exposed wiring from physical damage.

■ **Equipment** grounding or bonding conductors need to be included when you calculate conduit or tubing fill. The actual dimensions of the equipment grounding or bonding conductor need to be used in the calculation, regardless of whether the conductors are insulated or bare.

■ Conduit or tubing nipples that have a maximum length of 24 inches and are installed between boxes, cabinets, and similar enclosures can be filled to 60 percent of their total cross-sectional area, and **NEC** 310.15(B)(2)(a) adjustment factors do not have to be applied.

■ Multiconductor cables and similar conductors that are not included in Chapter 9 use the actual cable dimensions for all calculations.

■ Combinations of conductors of different sizes are listed in Table 5 and Table 5A for the dimensions of conductors and Table 4 for required conduit or tubing dimensions.

■ The maximum number of conductors that are all the same size and are allowed in a conduit or tubing is calculated using the next higher whole number when the calculation results in a decimal of 0.8 or larger.

■ Bare conductors that are permitted by other sections of the code are listed in Table 8.

■ Multiconductor cable or flexible cord with two or more conductors is treated as a single conductor when you are calculating for the percentage of conduit fill area. Cable that has elliptical cross sections is calculated using the major diameter of the ellipse as a circle diameter.

■ Table 8 on conductor properties provides a comprehensive, quick reference for conductor sizes, diameters, and types.

■ Table 9 is a detailed listing of AC resistance and reactance for 3-phase 600-volt cables that operate at 60 Hertz and have three single conductors in a conduit. Wire size and type examples are provided.

- Tables 12 (A) and 12 (B) show the required power-source limitations for power-limited fire-alarm systems. These fire-alarm circuits must be either inherently limited, which requires no overcurrent protection, or not inherently limited, which requires a combination of power source and overcurrent protection.

- Table B.310.7 lists various ampacities, by wire size, of three single insulated conductors that are rated up to 2000 volts, are installed in underground electrical ducts with three conductors per duct, and are based on an ambient temperature of 68 degrees F. Correction factors for other temperatures are included in the table.

- Table B.310.10 is based on three single insulated conductors that are rated up to 2000 volts and are direct-buried. You will notice that only cables approved to be directly buried are listed. The next several tables include the same type of information, based on varying conditions.

- Figures provided in is section, such as B.310.2 and B.310.3, illustrate cable-installation requirements based on burial depths and quantity of conductors.

- Annex A provides a listing of product safety standards.

- Annex B explains how to calculate and apply ampacity based on the number of conductors, the size of the wire, and the voltage of the system.

- Annex C lists the maximum number of conductors and fixture wires, all of permitted trade sizes of various conduits and tubing. The sizes are based on the total cross-sectional area including insulation. In Annex C you will find extensive conduit and tubing tables for conductors. For example, in Table C.1 you'll find that you are allowed to install three #8 AWG conductors in 1/2-inch EMT. Table C.1(A) provides similar information for compact conductors. If you run the EMT in a straight line, the interior diameter is 0.622 inches, but as soon as you include a bend in the conduit, the major internal diameter to the EMT can actually become slightly larger and therefore one of the conductors could slip between the other two. This creates the potential for a jam where the conductors exit the bend. The formula to calculate jam ratio is:

$$\text{Jam ratio} = \frac{\text{ID of the raceway}}{\text{OD of the conductor}}$$

To avoid a possible jam within the tubing, use the recommended jam ratios of 2.8 and 3.2. For various combinations of conductors of different sizes, you need to look at Tables 5 and 5.A, which provide the dimensions of conductors, and Table 4 for the applicable tubing or conduit dimensions.

- Annex D is a section of detailed examples using formulas for such applications as general lighting loads, the size of neutrals for feeders, demand factors, calculated loads for service conductors, and more. The examples vary from single-family units to multifamily dwellings, as well as electric range loads and motor-circuit calculations. Figure D9 illustrates an adjustable speed-drive control and how to determine the feeder ampacity. Figure D10 illustrates an adjustable speed-drive control.

- Annex E describes five different types of construction including fire-resistant and nonrated materials and provides two tables. Annex F is particularly handy, not only because is compares the location of code articles over the last three publications but because it briefly describes the articles so that you can jump to that section of the book if you are struggling to find an answer. Finally, Annex G explains how the code is administered and by whom as well as the issuing of permits, violations of the code, penalties, and the process for appeals.

CHAPTER

14

Code Locations and References for Additional Codes

Arc-fault Circuits in Bedrooms
NEC
210.12

Bathroom Branch Circuit
NEC
210.11 (C) (3)

Bonding of Metal Boxes
NEC
250.80
250.86
250.92 thru 250.106

Wire Sizes for Branch Circuits and Kitchen Appliance Circuits
NEC
210.19
210.21
210.50
210.52
Table 310.16

Conductor Fill for Boxes
NEC
314.16

Conductor Applications
NEC
366 Auxillary gutter
368 Busways
370 Cable bus
372 Cellular concrete floor raceways
374 Cellular metal floor raceways
376 Metal wireways
378 Nonmetallic raceways
380 Multioutlet assemblies
382 Nonmetallic extensions
384 Strut-type channel raceway
386 Surface metal raceway
388 Surface nonmetallic raceways

390 Underfloor raceways
392 Cable trays
394 Concealed knob and tube wiring
396 Messenger-supported wiring
398 Open wiring on insulators

Feeder Sizes

NEC
310.16

GFCI's in required locations

NEC
210.8

Ground Sizes and Locations

NEC
250.50 thru 250.66

Protection of Wires Run through Framing and Studs

NEC
300.21
300.4

Installation of Electrical Panel Clearances

NEC
408.18
110.26

Nipples with Locknuts and Bushings

NEC
300.4 B. (1) (F)

Receptacles on Wall Spaces and Counters

NEC
210.52(A) & (C)

Required number of Room Circuits & Outlets

NEC
210.12 & 210.19

Recessed Lighting Fixtures

NEC
410

Electrical wiring in Air Handling Plenums

NEC
300.22(B)

Protection of Electrical Wires and Cables in Attics

NEC
334.30

Washer and Dryer Circuits

NEC
220.16 (B) & 220.18

Referance Material

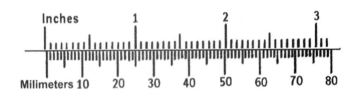

Use the tool above to convert Inches to Milimeters
Example 1/2 inch = 12mm "Milimeters"

APPENDIX FIGURE 1

Inches to milimeter conversion tool.

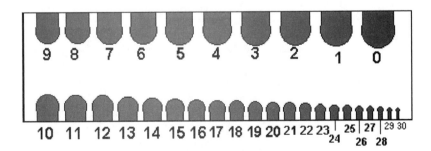

APPENDIX FIGURE 2

Wire guage.

Series Direct Current Circuits

◆ Total Resistance of a series circuit is equal to the sum of the individual resistances.

◆ The same current flows through each part of a series circuit.

◆ The total voltage across a series circuit is equal to the sum of the individual voltage drops.

◆ The voltage drop across a resistor in a series circuit is proportionate to the size of the resistor.

◆ The total power dissipated in a series circuit is equal to the sum of the individual power dissapation

---

Parallel Direct Current Circuits

The same voltage is present across each branch of a parallel circuit and is equal to the source voltage.

Current that flows through a branch of a parallel network is inversely proportional to the amount of resistance of the branch.

The total power loss in a parallel circuit is equal to the sum of the individual power dissapation

The total current of a parallel circuit is equal to the sum of the currents of the individual branches of the circuit.

The total resistance of a parallel circuit is equal to the shared sum of the reciprocals of the individual resistances of the circuit.

PARALLEL CIRCUIT RULES	
TOTAL VOLTAGE =	$E(1) = E(2) = E(3)$
TOTAL RESISTANCE =	$\dfrac{\text{VOLTS}}{\text{AMPERES}}$
VOLTS =	$\dfrac{\text{TOTAL VOLTAGE}}{\text{TOTAL AMPERES}}$

APPENDIX FIGURE 3

Basic electrical theory.

POWER IN SINGLE PHASE RESISTIVE CIRCUITS WITH A POWER FACTOR OF 100 PERCENT

◆To determine the power used by an individual resistor in a SERIES CIRCUIT USE THE FOLLOWING FORMULA:

$$POWER = I^2 \times R$$

◆To determine the power used by an individual resistor in a PARALLEL CIRCUIT USE THIS FORMULA:

$$POWER = \frac{E^2}{R}$$

◆To determine the total power used by an INDIVIDUAL CIRCUIT USE THIS FORMULA:

POWER = E (TOTAL VOLTAGE) x I (TOTAL CURRENT)

POWER IN ALTERNATING CURRENT CIRCUITS WHERE POWER FACTOR IS NOT 100 PERCENT

POWER = E x I x POWER FACTOR (FOR SINGLE PHASE)

POWER = E x I x 1.732 X POWER FACTOR (FOR THREE PHASE)

VOLT-AMPERES = E x I (FOR SINGLE PHASE)

VOLT-AMPERES = E x I x 1.732 (FOR THREE PHASE)

$$POWER\ FACTOR = \frac{TRUE\ POWER}{APPARENT\ POWER}$$

POWER CALCULATED IN THIS WAY IS CALLED TRUE POWER OR REAL POWER

APPARENT POWER FOUND BY CALCULATING VOLT-AMPERES.

APPENDIX FIGURE 4

Basic electrical theory.

MOTOR APPLICATION FORMULAS

HORSEPOWER = $\dfrac{1.732 \times \text{VOLTS} \times \text{AMPERES} \times \text{EFFICIENCY} \times \text{power factor}}{746}$
(for three phase motors)

THREE PHASE AMPERES = $\dfrac{746 \times \text{HORSEPOWER}}{1.732 \times \text{VOLTS} \times \text{EFFICIENCY} \times \text{POWER FACTOR}}$
(for three phase motors)

SYNCHRONOUS RPM = $\dfrac{\text{HERTZ} \times 120}{\text{NUMBER OF POLES}}$

	Maximum Horsepower for NEMA-Rated Motor Starters			
	Single-Phase		Three-Phase	
NEMA Size	115 Volt	230 Volt	208/230 Volt	460/575 Volt
00	1/3	1	1.5	2
0	1	2	3	5
1	2	3	7.5	10
2	3	7.5	10/15	25
3			25/30	50
4			40/50	100
5			75/100	200

APPENDIX FIGURE 5

Basic electrical theory.

Metric Conversion

Linear Measure

1 centimeter	0.3937 inch
1 inch	2.54 centimeters
1 decimeter	3.937 in., 0.328 foot
1 foot	3.048 decimeters
1 meter	39.37 inches, 1.0936 yds.
1 yard	0.9144 meter
1 dekameter	1.9884 rods
1 rod	0.5029 dekameter
1 kilometer	0.62137 mile
1 mile	1.6094 kilometers

Square Measure

1 sq. centimeter	0.1550 sq. inches
1 sq. inch	6.452 sq. centimeters
1 sq. decimeter	0.1076 sq. foot
1 sq. foot	9.2903 sq. decimeters
1 sq. meter	1.196 yards
1 sq. yard	0.8361 sq. meter
1 hectare	2.471 acres
1 acre	0.4047 hectare
1 sq. kilometer	0.386 sq. mile

Measure of Volume

1 cu. centimeter	0.06 1 cu. inch
1 cu. Inch	16.39 cu. centimeters
1 cu. decimeter	0.0353 cu. foot
1 cu. foot	28.3 17 cu. decimeters
1 cu. yard	0.7646 cu. meters
1 cu. meter	0.2759 cord
1 cord	3.625 steres
1 liter	0.908 qt. dry 1.0567 qts. liq.
1 quart dry	1.101 liters
1 quart liquid	0.9463 liter
1 dekaliter	2.6417 gals, 1.135 pks.
1 gallon	0.3785 dekaliter
1 peck	0.881 dekaliter
1 hectoliter	2.8378 bushels
1 bushel	0.3524 hectoliter

Weights

1 gram	0.03527 ounce
1 ounce	28.35 grams
1 kilogram	2.2046 pounds
1 pound	0.4536 kilogram

APPENDIX FIGURE 6

Metric equivalents.

◆Conversion Formulas

Busbar Ampacity AL = 700A Sq. in. and CU = 1000A Sq. in.

Centimeters = Inches x 2.54

Inch = 0.0254 Meters

Inch = 2.54 Centimeters

Inch = 25.4 Millimeters

Kilometer = 0.6213 Miles

Meter = 39.37 Inches

Millimeter = 0.03937 Inch

Temp C = (Temp F - 32)/1.8

Temp F = (Temp C x 1.8) + 32

Yard = 0.9144 Meters

Length of Coiled Wire =
   Diameter of Coil (average) x Number of Coils x π

APPENDIX FIGURE 7

Conversion formulas.

◆Electrical Formulas Based on 60 Hz

Effective (RMS) AC Amperes = Peak Amperes x 0.707

Effective (RMS) AC Volts = Peak Volts x 0.707

Efficiency = Output/Input

Horsepower = Output Watts/746

Input = Output/Efficiency

Neutral Current (Wye) = $\sqrt{A^2 + B^2 + C^2 - (AB + BC + AC)}$

Output = Input x Efficiency

Peak AC Volts = Effective (RMS) AC Volts x $\sqrt{2}$

Peak Amperes = Effective (RMS) Amperes x $\sqrt{2}$

Power Factor (PF) = Watts/VA

VA (apparent power) = Volts x Ampere or Watts/Power Factor

VA 1-Phase = Volts x Amperes

VA 3-Phase = Volts x Amperes x $\sqrt{3}$

Watts - Single-Phase = Volts x Amperes x Power Factor

Watts - Three-Phase = Volts x Amperes x Power Factor x $\sqrt{3}$

APPENDIX FIGURE 8

Electrical formulas.

◆CURRENT FLOW & CIRCUITS

◆Parallel Circuits

~Total resistance is always less than the smallest resistor

$$RT = 1/(1/R1 + 1/R2 + 1/R3 \text{ (etc)}$$

~Total current is equal to the sum of the currents of all parallel resistors

~Total power is equal to the sum of power of all parallel resistors

~Voltage is the same across each of the parallel resistors

◆Series Circuits

~Total resistance is equal to the sum of all the resistors

~Current in the circuit remains the same through all the resistors

~Voltage source is equal to the sum of voltage drops of all resistors

~Power of the circuit is equal to the sum of the power of all resistors

◆Voltage Drop

VD (1-Phase) = 2KID/CM

VD (3-Phase) = $\sqrt{3}$ KID/CM

CM (1-Phase) = 2KID/VD

CM (3-Phase) = $\sqrt{3}$ KID/VD

APPENDIX FIGURE 9

Electrical formulas.

◆Transformer Calculations

Secondary Amperes 1-Phase = VA/Volts

Secondary Amperes 3-Phase = VA/Volts x $\sqrt{3}$

Secondary Available Fault 1-Phase = VA/(Volts x %impedance)

Secondary Available Fault 3-Phase = VA/(Volts x $\sqrt{3}$ x %Impedance)

Delta 4-Wire: Line Amperes = Phase (one winding) Amperes x $\sqrt{3}$

Delta 4-Wire: Line Volts = Phase (one Winding) Volts

Delta 4-Wire: High-Leg Voltage (L-to-G) = Phase (one winding) Volts x 0.5 x $\sqrt{3}$

Wye: Line Volts = Phase (one winding) Volts x $\sqrt{3}$

Wye: Line Amperes = Phase (one winding) Amperes

◆Voltage and Current:

Primary (p) secondary (s) Power(p) = power (s) or Ep x Ip = Es x Is

•Ep =	$\dfrac{\text{Es x Is}}{\text{Ip}}$
•Ip =	$\dfrac{\text{Es x Is}}{\text{Ep}}$
•Is =	$\dfrac{\text{Ep x Ip}}{\text{Es}}$
•Es =	$\dfrac{\text{Ep x Ip}}{\text{Is}}$

Voltage and Turns in Coil:
Voltage (p) x Turns (s) = Voltage (s) x Turns (p) or Ep x Ts = Es x Ip

•Ep =	$\dfrac{\text{Es x Ip}}{\text{Ts}}$
• Ts =	$\dfrac{\text{Es x Tp}}{\text{Ep}}$
•Tp =	$\dfrac{\text{Ep x Ts}}{\text{Es}}$
• Es =	$\dfrac{\text{Ep x Ts}}{\text{Tp}}$

APPENDIX FIGURE 10

Electrical formulas.

VOLTAGE DROP

Voltage is contained in a short conductor because the power flow meets less resistance

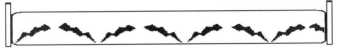

Voltage spreads out in a longer conductor and the resistance to the power flow increases and the voltage drops

◆ Voltage Drop Formulas

Single Phase (2 or 3 wire)	$VD = \dfrac{2 \times K \times I \times L}{CM}$	**K** = ohms per mil foot **(Copper = 12.9 at 75°)** **(Alum = 21.2 at 75°)** *Note:* **K** *value changes with temperature. See NEC Chapter 9, Table 8* **L** = Length of conductor in feet **I** = Current in conductor (amperes) **CM** = Circular mil area of conductor

◆ AC/DC Formulas

E = Voltage / I = Amps /W = Watts / PF = Power Factor / Eff = Efficiency / HP = Horsepower

To Find	Direct Current	AC / 1phase 115v or 120v	AC / 1phase 208,230, or 240v	AC 3 phase All Voltages
Amps when Horsepower is Known	$\dfrac{HP \times 746}{E \times Eff}$	$\dfrac{HP \times 746}{E \times Eff \times PF}$	$\dfrac{HP \times 746}{E \times Eff \times PF}$	$\dfrac{HP \times 746}{1.73 \times E \times Eff \times PF}$
Amps when Kilowatts is known	$\dfrac{kW \times 1000}{E}$	$\dfrac{kW \times 1000}{E \times PF}$	$\dfrac{kW \times 1000}{E \times PF}$	$\dfrac{kW \times 1000}{1.73 \times E \times PF}$
Amps when kVA is known		$\dfrac{kVA \times 1000}{E}$	$\dfrac{kVA \times 1000}{E}$	$\dfrac{kVA \times 1000}{1.73 \times E}$
Kilowatts	$\dfrac{I \times E}{1000}$	$\dfrac{I \times E \times PF}{1000}$	$\dfrac{I \times E \times PF}{1000}$	$\dfrac{I \times E \times 1.73\ PF}{1000}$
Kilovolt-Amps		$\dfrac{I \times E}{1000}$	$\dfrac{I \times E}{1000}$	$\dfrac{I \times E \times 1.73}{1000}$
Horsepower (output)	$\dfrac{I \times E \times Eff}{746}$	$\dfrac{I \times E \times Eff \times PF}{746}$	$\dfrac{I \times E \times Eff \times PF}{746}$	$\dfrac{I \times E \times Eff \times 1.73 \times PF}{746}$

◆ Power - DC Circuits

Watts = E x I
Amps = W / E

◆ AC Efficiency and Power Factor Formulas

E = Voltage / I = Amps /W = Watts / PF = Power Factor / Eff = Efficiency / HP = Horsepower

To Find:	Single Phase	Three Phase
Efficiency	$\dfrac{746 \times HP}{E \times I \times PF}$	$\dfrac{746 \times HP}{E \times I \times PF \times 1.732}$
Power Factor	$\dfrac{Input\ Watts}{V \times A}$	$\dfrac{Input\ Watts}{E \times I \times 1.732}$

APPENDIX FIGURE 11

Electrical formulas.

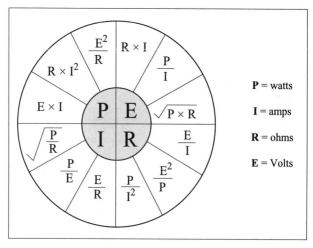

APPENDIX FIGURE 12

Ohm's law.

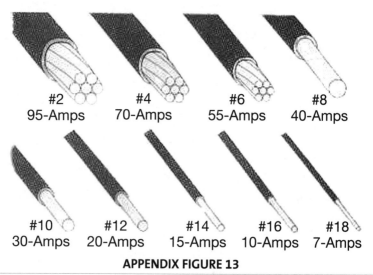

APPENDIX FIGURE 13

Cable sizes and amperages.

COMMON ELECTRICAL TERMS

General Terms

Adapter An accessory used for interconnecting non-mating devices or converting an existing device for modified use.

Ballast A transformer that steps down AC line voltage to voltage that can be used by fluorescent or other types of lighting. Ballast's may be electromagnetic or electronic

Cord Connector A portable receptacle designed for attachment to or provided with flexible cord, not intended for fixed mounting.

Flanged Inlet A plug intended for flush mounting on appliances or equipment to provide a means for power connection via a cord connector.

Flanged Outlet A receptacle intended for flush mounting on appliances or equipment to provide a means for power connection via an inserted plug.

Fluorescent Starter A device with a voltage-sensitive switch and a capacitor that provides a high-voltage pulse to start a fluorescent lamp. Rated in watts.

Lampholder A device with contacts that establishes mechanical and electrical connection to an inserted lamp.

Plug A device with male contacts intended for insertion into a receptacle to establish electrical connection between the attached flexible cord and the conductors connected to the receptacle.

Receptacle A device with female contacts designed for fixed installation in a structure or piece of equipment and which is intended to establish electrical connection with an inserted plug.

Switch A device for making, breaking, or changing the connections in an electric circuit.

Wallplate A plate designed to enclose an electrical box, with or without a device installed within the box.

APPENDIX FIGURE 14

Electrical terms.

Lampholders

<u>Bayonet</u> Designed for incandescent lamps having an unthreaded metal shell with two diametrically opposite keyways that mate with the keyways on the lampholder. Pushing down on the bulb and turning it clockwise in the lampholder locks the bulb in place.

<u>Candelabra</u> A small screw-base threaded lampholder designed for candelabra-base incandescent lamps commonly used in chandeliers, night lights, and ornamental lighting.

<u>Circline</u> A four-contact, double-ended lampholder designed for use with tubular, circular fluorescent lamps.

<u>Compact Fluorescent</u> A lampholder designed for the Compact Fluorescent Lamps (CFL's) that are increasingly being used to replace incandescent lamps for energy efficiency.

<u>Dimmer</u> An electronic device with either a round knob, slide lever or finger-tip controlled buttons used to dim/brighten incandescent lighting. Available in a variety of wattages; fluorescent version also available.

<u>Double-Contact Recessed</u> Designed for high-output fluorescent lamps.

<u>Edison Base</u> An internally-threaded lampholder, with the inner shell approx. 1" in diameter. Designed for widely-used standard medium base lamps.

<u>Electrolier</u> Similar to the Edison Medium Base lampholder, but with a smaller outer diameter. Incandescent Designed for use with all manufactured incandescent lamps, most of which have threaded bases.

<u>Key</u> A lampholder with a flat or round "key" knob that operates an internal switching mechanism ("Keyless" lampholders do not provide an internal switching mechanism).

<u>Lumiline</u> A specially designed lampholder for tubular Lumiline-type incandescent lamps, typically used in bathrooms and retail display cases.

<u>Mogul</u> The largest screw-in type lampholder, designed for mogul incandescent lamps with a screw base of approx. 1 1/2" dia. Used in street lights and numerous commercial/industrial applications. Medium Bi-Pin A fluorescent lampholder with two contacts, used in pairs. For type T-8 tubular fluorescent lamps, approx. 1" in diameter.

<u>Miniature Bi-Pin</u> Similar to medium bi-pin lampholders, but designed for type T-5 tubular fluorescent lamps, approx. 5/8" in diameter.

<u>Outlet Box</u> Medium-base incandescent lampholder designed for mounting in 3 1/4" or 4" electrical boxes. Available with or without pull-chain mechanism, and with or without built-in receptacle. Pull-Chain An incandescent lampholder with an internal switching mechanism that is activated by pulling down on a beaded chain or cord.

<u>Push-Through</u> An incandescent lampholder with an insulated lever that is pushed from either side to activate an internal ON/OFF switching mechanism.

<u>Slimline</u> Single-Pin A fluorescent lampholder with a single contact designed for Slimline fluorescent lamps such as the T-12 (1 1/2" dia.), T-8 (1" dia.), and the smaller version T-6 (3/4" dia.).

<u>Snap-In</u> An incandescent or compact fluorescent lampholder with factory-assembled spring clips that securely snap into a panel cutout without requiring additional fasteners. Surface-Mounted A lampholder of any type that mounts on a flat or plane surface.

APPENDIX FIGURE 15

Electrical terms.

Receptacles

AL/CU 30A, 50A or 60A receptacles designated for use with aluminum or copper circuit conductors, identified by "AL/CU" stamped on the device. Receptacles without this designation must never be used with aluminum circuit conductors.

Clock Hanger A single, recessed receptacle with a specialized cover plate that provides a hook or other means of supporting a wall clock.

CO/ALR 15A or 20A receptacles designated for use with aluminum or copper circuit conductors, identified by "CO/ALR" stamped on the device. Receptacles without this designation must never be used with aluminum circuit conductors.

Corrosion Resistant A receptacle constructed of special materials and/or suitably plated metal parts that is designed to withstand corrosive environments. Corrosion resistant devices must pass the ASTM B117-13 five-hundred hour Salt Spray (Fog) Test with no visible corrosion.

Display Receptacle with a special cover plate intended for flush mounting on raised floors or walls.

Duplex Two receptacles built with a common body and mounting means; accepts two plugs.

Explosion Proof A receptacle constructed to meet the requirements of hazardous locations as defined by the National Electrical Code, NFPA-70.

Four-In-One or "Quad" A receptacle in a common housing that accepts up to four plugs. Four-In-One receptacles can be installed in place of duplex receptacles mounted in a single-gang box, providing a convenient means of adding receptacles without rewiring.

GFCI (Ground Fault Circuit Interrupter) A receptacle with a built in circuit that will detect leakage current to ground on the load side of the device. When the GFCI detects leakage current to ground, it will interrupt power to the load side of the device, preventing a hazardous ground fault condition. GFCI receptacles must conform to UL Standard 943 Class A requirements, and their use is required by the National Electric Code NFPA-70 in a variety of indoor and outdoor locations.

Hospital Grade A receptacle designed to meet the performance requirements of high-abuse areas typically found in health care facilities. These receptacles are tested to the Hospital Grade requirements of Underwriters Laboratories Inc. Standard 498.

Interchangeable A receptacle or combination of receptacles with a common mounting dimension that may be installed on a single or multiple-opening mounting strap.

Isolated Ground Receptacles intended for use in an Isolated Grounding system where the ground path is isolated from the facility grounding system. The grounding connection on these receptacles is isolated from the mounting strap.

Lighted (Illuminated) A receptacle with a face that becomes illuminated when the device is connected to an energized electrical circuit.

Locking A receptacle designed to lock an inserted plug with a matching blade configuration when the plug is rotated in a clockwise direction. The plug can only be removed by first turning it in a counter-clockwise direction.

Safety or Tamper-Resistant A receptacle specially constructed so that access to its energized contacts is limited. Tamper-resistant receptacles are required by the National Electric Code NFPA-70 in specific pediatric care areas in health care facilities.

Single A receptacle that accepts only one plug.

Split-Circuit A duplex receptacle that allows each receptacle to be wired to separate circuits. Most duplex receptacles provide break-off tabs that allow them to be converted into split-circuit receptacles.

Straight Blade A non-locking receptacle into which mating plugs are inserted at a right angle to the plane of the receptacle face.

Surface-Mounted Any receptacle that mounts on a flat or plane surface.

Surge-Suppression A receptacle with built-in circuitry designed to protect its load side from high-voltage transients and surges. The circuitry will limit transient voltage peaks to help protect sensitive electronic equipment such as PC's, modems, audio/video equipment, etc.

Triplex A receptacle with a common mounting means which accepts three plugs.

Weatherproof A receptacle specially constructed so that exposure to weather will not interfere with its operation.

APPENDIX FIGURE 16

Electrical terms.

Wallplates

Combination A multiple- gang wallplate with openings in each gang to accommodate different devices.

Flush A wallplate designed for flush-mounting with wall surfaces or the plane surfaces of electrical equipment.

Gang A term that describes the number of devices a wallplate is sized to fit (i.e. "2- gang" designates two devices).

Midway Wallplates that are approx. 3/8" higher and wider than the standard size that can be mounted onto larger volume outlet boxes and/or used to hide wall surface irregularities. These wallplates are approx. 1/4" deep to ensure a proper fit when used with protruding devices. Oversized Wallplates that are approx. 3/4" higher and wider than the standard size and are used to conceal greater wall irregularities than those hidden by Midway wallplates. These wallplates are approx. 1/4" deep to ensure a proper fit when used with protruding devices.

Modular Individual-section wallplates with different openings that can be configured into a multi-gang plate.

Multi-Gang A wallplate that has two or more gangs.

Tandem A wallplate with individual gangs arranged vertically one above the other.

Weatherproof (with Cover Closed) A UL Listed cover that meets specific test standards for use in wet and damp locations with the cover closed.

Weatherproof (with Cover Open) A UL Listed cover that meets specific test standards for use in wet and damp locations with the cover open or closed.

APPENDIX FIGURE 17

Electrical terms.

Switches

AC/DC A switch designated for use with either Alternating Current (AC) or Direct Current (DC).

AC Only A switch designated for use with Alternating Current (AC) only.

Dimmer A switch with electronic circuitry that provides DIM/BRIGHT control of lighting loads.
Door A momentary contact switch, usually installed on a doorjamb, that is activated when the door is opened or closed.

Double-Pole, Single-Throw (DPST) A switch that makes or breaks the connection of two circuit conductors in a single branch circuit. This switch has four terminal screws and ON/OFF markings.

Double-Pole, Double-Throw (DPDT) A switch that makes or breaks the connection of two conductors to two separate circuits. This switch has six terminal screws and is available in both momentary and maintained contact versions, and may also have a center OFF position.

Feed-Through An in-line switch that can be attached at any point on a length of flexible cord to provide switching control of attached equipment.

APPENDIX FIGURE 18

Electrical terms.

Flush-Mounted A switch designed for flush installation with the surface of a panel or equipment.

Four-Way A switch used in conjunction with two 3-Way switches to control a single load (such as a light fixture) from three or more locations. This switch has four terminal screws and no ON/OFF marking.

Horsepower Rated A switch with a marked horsepower rating, intended for use in switching motor loads.

Interchangeable A switch or combination of switches with a common mounting dimension that may be installed on a single or multiple-opening mounting strap.

Lighted Handle A switch with an integral lamp in its actuator (toggle, rocker or pushbutton) that illuminates when the switch is connected to an energized circuit and the actuator is in the OFF position.

Low-Voltage A switch rated for use on low-voltage circuits of 50 volts or less.

L-Rated A switch specially designated with the letter "L" in its rating that is rated for controlling tungsten filament lamps on AC circuits only.

Maintained Contact A switch where the actuator (toggle, rocker, pushbutton or key mechanism) makes and retains circuit contact when moved to the ON position. The contacts will only be opened when the actuator is manually moved to the OFF position. Ordinary light switches are maintained contact switches.

Manual Motor Controller A switch designed for controlling small DC or AC motor loads, without overload protection.

Mercury A type of switch that uses mercury as the contact means for making and breaking an electrical circuit.

Momentary Contact A switch that makes circuit contact only as long as the actuator (toggle, rocker, pushbutton or key mechanism) is held in the ON position, after which it returns automatically to the OFF position. This is a "Normally Open" switch. A "Normally Closed" switch will break circuit contact as long as it is held in the OFF position, and then automatically return to the ON position. Available in "Center OFF" versions with both Momentary ON and Momentary OFF positions.

Pendant A type of switch designed for installation at the end of a length of portable cord or cable.

Pilot Light A switch with an integral lamp in its actuator (toggle, rocker or pushbutton) that illuminates when the switch is connected to an energized circuit and the actuator is in the ON position.

Pull A switch where the making or breaking of contacts is controlled by pulling downward or outward on the actuator mechanism.

Push Button A switch with an actuator mechanism that is operated by depressing a button.

Rotary A switch where rotating the actuator in a clockwise direction makes the circuit connection, and then rotating the actuator in either the same or opposite direction breaks the connection.

Single-Pole, Double-Throw (SPDT) A switch that makes or breaks the connection of a single conductor with either of two other single conductors. This switch has 3 terminal screws, and is commonly used in pairs and called a "Three-Way" switch.

Single-Pole, Single-Throw (SPST) A switch that makes or breaks the connection of a single conductor in a single branch circuit. This switch has two screw terminals and ON/OFF designations. It is commonly referred to as a "Single-Pole" Switch.

Slide A switch with a slide-action actuator for making or breaking circuit contact. Dimmer switches and fan speed controls are also available with slide-action mechanisms for lighting and fan speed control.

Surface-Mounted Any switch that mounts on a flat or plane surface.

Three-Position, Center OFF A two circuit switch, either maintained or momentary contact, where the OFF position is designated as the center position of the actuator.

Three-Way A switch, always used in pairs, that controls a single load such as a light fixture from two locations. This switch has three terminal screws and has no ON/OFF marking.

Time Delay A switch with an integral mechanism or electronic circuit that will automatically switch a load OFF at a predetermined time interval.

Timer A switch with an integral mechanism or electronic circuit that can be set to switch an electrical load ON at a predetermined time.

Toggle A switch with a lever-type actuator that makes or breaks switch contact as its position is changed.

T-Rated A switch specially designated with the letter "T" in its rating that is rated for controlling tungsten filament lamps on direct current (DC) or alternating current (AC) circuits.

APPENDIX FIGURE 18 *(Continued)*

Electrical terms.

Surge Suppression

Clamping Voltage The peak voltage that can be measured after a Surge Protective Device has limited or "clamped" a transient voltage surge. Clamping voltage must be determined by using IEEE Standard C62 testing and evaluated by UL Standard 1449.

Joule Rating The measurement of a Surge Protective Device's ability to absorb heat energy created by transient surges. Note that the Joule rating is not a part of IEEE or UL Standards. It is not as significant a specification as Clamping Voltage, Maximum Surge Current and other parameters recognized by these agencies.

Transient Voltage Surges High-speed, high-energy electrical disturbances present on AC power lines and data and communication lines, generated by utility switching, motor-load switching and lightning strikes.

Response Time The interval of time it takes for a surge protective device to react to a transient voltage surge. Note that this parameter is not a part of IEEE or UL Standards and is only based on estimations made by manufacturers.

Surge Protective Device See "Transient Voltage Surge Suppressor (TVSS)" definition. Transient Voltage Surge Suppressor (TVSS) A device designed to protect sensitive electronic equipment such as computers and computer peripherals, logic controls, audio/video equipment and a wide range of microprocessor-based (computer chip) equipment from the harmful effects of transient voltage surges. Also referred to as a Surge Protective Device (SPD).

Maximum (Peak) Surge Current The peak surge current a Surge Protective Device can withstand, based on IEEE Standard C62.45 test waveforms.

MOV (Metal Oxide Varistor) The primary component used in most Surge Protective Devices to clamp down transient voltages.

UL 1449 Listing The industry standard for Surge Protective Devices. A Surge Protective Device must have a UL 1449 Surge Suppression rating on its label in order to verify that the device has been tested with IEEE standardized waveforms. Devices without this identification should not be considered reliable surge protective devices.

APPENDIX FIGURE 19

Electrical terms.

SINGLE-PHASE

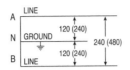

120/240 V, 3 W (240/480 V, 3 W)
THREE-WIRE

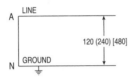

120 V, 2 W (240 V, 2 W) [480 V, 2 W]
TWO-WIRE

POLYPHASE

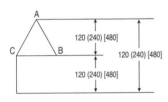

120 V, 3 W (240 V, 3 W) [480 V, 3 W]
THREE-PHASE, THREE-WIRE

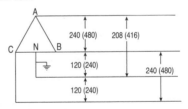

240/120 V, 4 W (480/240 V, 4 W)
THREE-PHASE, FOUR-WIRE DELTA

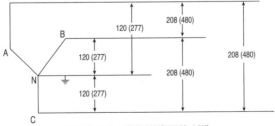

208Y/120 V, 4 W (480Y/277 V, 4 W)
THREE-PHASE, FOUR-WIRE WYE

APPENDIX FIGURE 20

Standard voltages.

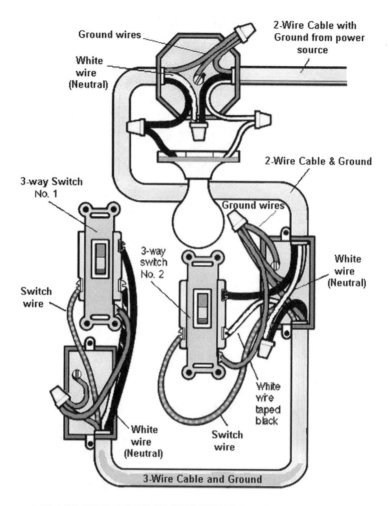

Ground wires

White wire (Neutral)

2-Wire Cable with Ground from power source

3-way Switch No. 1

2-Wire Cable & Ground

Ground wires

3-way switch No. 2

White wire (Neutral)

Switch wire

White wire taped black

White wire (Neutral)

Switch wire

3-Wire Cable and Ground

LUMINAIRE WIRING CONNECTION ON 3-WAY SWITCHES

APPENDIX FIGURE 21

Light wiring diagram.

AC SYSTEM GROUNDING ELECTRODE CONDUCTOR RATINGS
And
Derived conductors of Separately Derived AC Systems

COPPER	ALUMINIMUM OR COPPER-CLAD ALUMINMUM	COPPER	ALUMINIMUM OR COPPER-CLAD ALUMINMUM
Size of Largest Ungrounded Service Entrance Conductors or Equivalent Area of Parallel Conductors	Size of Largest Ungrounded Service Entrance Conductors or Equivalent Area of Parallel Conductors	Size of AWG Grounding Electrode	Size of AWG Grounding Electrode
2 or smaller	1/0 or smaller	#8 AWG	#6 AWG
1 OR 1/0	2/0 or 3/0	#6 AWG	#4 AWG
2/0 or 3/0	4/0 OR 250	#4 AWG	#2 AWG
Over 3/0 to 350	Over 250 to 500	#2 AWG	1/0
Over 350 to 600	Over 500 to 900	1/0	3/0
Over 600 to 1100	Over 900 to 1750	2/0	4/0
Over 1100	Over 1750	3/0	250

APPENDIX FIGURE 22

Grounding electrode.

MINIMUM CONDUCTOR SIZES

CONDUCTOR VOLTAGE	Copper	Aluminum or Copper-Clad Aluminum
0 - 2000 VOLTS	#14 AWG	#12 AWG
2100 - 4000 VOLTS 4001 - 8000 VOLTS	#8 AWG	#8 AWG
8001 - 10,000 VOLTS 10,001 - 15,000 VOLTS	#2 AWG	#2 AWG
15,001 - 25,000 VOLTS 25,001 - 28,000 VOLTS	#1 AWG	#1 AWG
28,001 - 35,000 VOLTS	1/0 AWG	1/0 AWG

APPENDIX FIGURE 23

Minumum conductor sizes.

NEMA RATING FOR ENCLOSURES

NEMA and other organizations have established standards of enclosure construction for control equipment. In general, equipment would be enclosed for one or more of the following reasons:

1. Prevent accidental contact with live parts.
2. Protect the control from harmful environmental conditions.
3. Prevent explosion or fires which might result from the electrical arc caused by the control.

Common types of enclosures per NEMA classification numbers are:

NEMA I - GENERAL PURPOSE : The general purpose enclosure is intended primarily to prevent accidental contact with the enclosed apparatus. It is suitable for general purpose applications indoors where it is not exposed to unusual service conditions. A NEMA I enclosure serves as protection against dust and light indirect splashing, but is not dusttight.

NEMA 3 - DUSTTIGHT, RAINTIGHT : This enclosure is intended to provide suitable protection against specified weather hazards. A NEMA 3 enclosure is suitable for application outdoors, on ship docks, canal and construction work, and for application in subways and tunnels. It is also sleet-resistant.

NEMA 3R - RAINPROOF, SLEET RESISTANT : This enclosure protects against interference in operation of the contained equipment due to rain, and resists damage from exposure to sleet. It is designed with conduit hubs and external mounting, as well as drainage provisions.

NEMA 4 - WATERTIGHT : A watertight enclosure is designed to meet the hose test described in the following note: "Enclosures shall be tested by subjection to a stream of water. A hose with a one inch nozzle shall be used and shall deliver at least 65 gallons per minute. The water shall be directed on the enclosure from a distance of not less than 10 feet and for a period of five minutes. During this period it may be directed in any one or more directions as desired. There shall be no leakage of water into the enclosure under these conditions."

A NEMA 4 enclosure is suitable for applications outdoors on ship docks and in dairies, breweries, etc.

NEMA 4X - WATERTIGHT, CORROSION-RESISTANT : These enclosures are generally constructed along the lines of NEMA 4 enclosures except they are made of a material that is highly resistant to corrosion. For this reason, they are ideal in applications such as paper mills, meat packing, fertilizer and chemical plants where contaminants would ordinarily destroy a steel enclosure over a period of time.

NEMA 7 - HAZARDOUS LOCATIONS - CLASS I : These enclosures are designed to meet the application requirements of the National Electrical Code for Class I hazardous locations. In this type of equipment, the circuit interruption occurs in air.

"Class I locations are those in which flammable gases or vapors are or may be present in the air in quantities sufficient to produce explosive or ignitable mixtures."

NEMA 9 HAZARDOUS LOCATIONS - CLASS II : These enclosures are designed to meet the application requirements of the National Electrical Code for Class II hazardous locations. Class II locations are those which are hazardous because of the presence of combustible dust.

The letter or letters following the type number indicates the particular group or groups of hazardous locations (as defined in the National Electrical Code) for which the enclosure is designed. The designation is incomplete without a suffix letter or letters.

NEMA 12 - INDUSTRIAL USE : The NEMA 12 enclosure is designed for use in those industries where it is desired to exclude such materials as dust, lint, fibers and flyings, oil see page or coolant see page. There are no conduit openings or knockouts in the enclosure, and mounting is by means of flanges or mounting feet.

NEMA 13 - OILTIGHT, DUSTTIGHT : NEMA 13 enclosures are generally of cast construction, gasketed to permit use in the same environments as NEMA 12 devices. The essential difference is that, due to its cast housing, a conduit entry is provided as an integral part of the NEMA 13 enclosure, and mounting is by means of blind holes, rather than mounting brackets.

APPENDIX FIGURE 24
NEMA rated enclosures.

NFPA 70E

✗ The current edition of the **National Fire Protection Association 70E** was approved as an American National Standard on February 11, 2004 and is overseen by the Committee on Electrical Safety Requirements for Employee Workplaces.

✗ **NFPA 70E** covers a full range of electrical safety issues, including safety-related work practices, special equipment requirements, installations, and arc-flash and arc-blast prevention.

✗ OSHA now bases its electrical safety mandates, which are generally addressed in OSHA 1910 Subpart S and OSHA 1926 Subpart K, on the comprehensive information provided by **NFPA 70E**.

✗ Important "recommendation" standards include the following:

- Electrical Hazard Analysis for Shock and Flash
- Energized Electrical Work Permit mandates and samples
- Approach Boundaries for Shock
- Arc-flash Boundaries
- Personal Protection Equipment requirements for electrical safety

APPENDIX FIGURE 25

NFPA 70E.

◆Quick Code References

AC-DC: Panelboards & Ratings - SECTION [408.36 (C)]

Quick Reference List

Air Conditioners & Refrigeration Equipment - Article 440

Appliances: Load Calculations - SECTION [220.14], [220.40] Table [220.55]

Audio Signal Processing: Grounding - SECTION [640.7]

Bathroom Wiring: Branch Circuits -SECTION [210.11 (C) (3); Receptacles SECTION [210.8 (A)(1)

Branch Circuits: Articles 210 & 220

Breakers and Fuse Ratings: SECTION [240.6(A)]

Conductor Ampacity: SECTION [310.15] and Table [310.16]

Conductors: Number (quantity) Allowed - Chapter 9, Table 1 - Annex C

Equipment Grounding Conductors: SECTION [250.122]

Grounding Electrode Conductors: SECTION [250.66]

Motor Conductor Sizing: SECTION [430.22] for Single and SECTION [430.24] for Multiple

Feeders: Calculation of loads - SECTION [215.2(A), [220.40]

Fire Alarm Systems: Conductors: SECTION [760.27]

Flexible Metal Conduit: Number of Conductors: SECTION [348.22]

Generators: Article 445

Hazardous Locations: Article 500

Loads: Calculations: Article 220, Annex D

Motor Short-Circuit Protection: SECTION [430.52]

Panelboards: Distribution - SECTIONS 550, 551, 552

Service Entrance Conductors: Article 230

Transformer Overcurrent Protection: SECTION [450.3]

APPENDIX FIGURE 26

Code locations.

Index